Formation of a Digital Image

The Imaging Chain Simplified

Formation of a Digital Image

The Imaging Chain Simplified

Robert D. Fiete

SPIE
PRESS
Bellingham, Washington USA

Library of Congress Cataloging-in-Publication Data

Fiete, Robert D.
Formation of a digital image : the imaging chain simplified / Robert D. Fiete.
p. cm.
Includes bibliographical references and index.
ISBN 978-0-8194-8976-0
1. Photographic optics--Mathematics. 2. Digital cameras--Mathematical models. 3. Photography--Digital techniques. I. Title.
TR220.F539 2012
771.3--dc23

2011046487

Published by
SPIE
P.O. Box 10
Bellingham, Washington 98227-0010 USA
Phone: +1 360.676.3290
Fax: +1 360.647.1445
Email: Books@spie.org
Web: http://spie.org

Printed in the United States of America.
First printing

To my family and friends, who don't understand my obsession with this thing called the imaging chain.

Contents

Preface

Many of us working in technical fields hesitate when people ask us to explain what we do for a living. After I received my degree in optics, I sometimes found myself saying "yes" when older relatives asked if I worked on eyeglasses. Explaining that I work on the design of digital cameras by modeling the imaging chain is sure to produce blank stares or questions such as, "Can I buy an imaging chain in the store?" When I attempt to explain further, I usually fall into the trap of using the scientific language that I have grown accustomed to with my colleagues. This, of course, produces a long yawn from the listener, who then says, "Interesting," and saunters into another room.

Formation of a Digital Image: The Imaging Chain Simplified discusses the process that creates digital images for people who don't want to be blinded with equations and bored with geek speak. This book is written for individuals who work with camera designers and want to know but are sometimes afraid to ask why they keep babbling about an MTF or some other mysterious acronym. If you ever wonder why pinstripe suits turn psychedelic on TV or why crosses appear on pictures of stars, I hope you will find this book helpful, and I apologize in advance for sometimes getting too close to geek speak. Habits are very hard to break. If nothing else, this book has photos of puppies that you will hopefully enjoy.

Bob Fiete
January 2012
Rochester, New York

Chapter 1
Making Pictures

1.1 The Magical Process

When most of us pick up our digital camera and push the button, we rarely put any thought into the "magic" that creates the photograph or digital image. Many people in the early days of photography refused to have their picture taken because they thought this magic of capturing their image was evil. What was once considered evil magic is simply a process that is well understood by image scientists. So, what actually happens in that moment, typically a fraction of a second, when a picture is created? Let's find out!

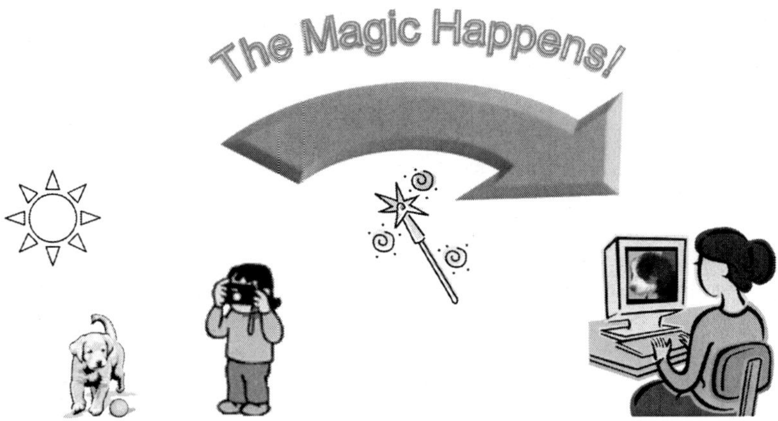

What is this process that creates a digital picture?

The step-by-step process that creates the picture of that adorable puppy for us to cherish forever is called the imaging chain. Scientists and engineers study and analyze the imaging chain, sometimes called the imaging pipeline, to help them make smaller, less-expensive cameras that produce better image quality. The imaging chain is typically divided into major links, and each link is studied by experts to understand its part in the creation of an image. These links are typically the light, camera, processing, and way we display and interpret information in the image. The camera is usually further divided into two individual links, the optics and sensor. These are thought of as separate links, because these elements of the camera are complicated enough on their own, and each has its own significant challenges that must be overcome to make a good photograph.

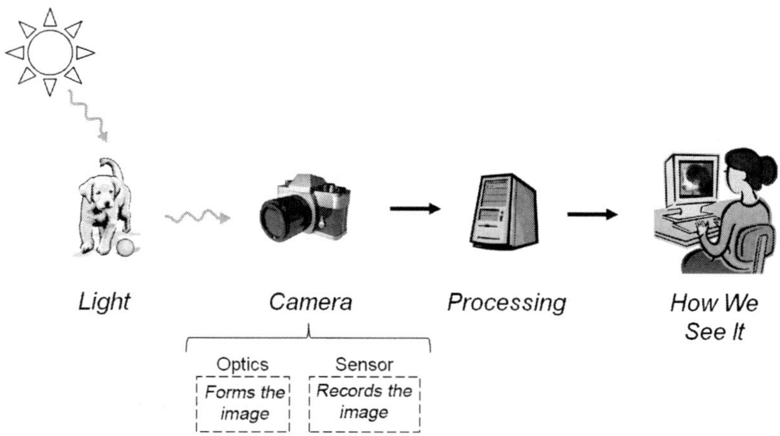

The links of an imaging chain make up the magical process.

Because individual links of the imaging chain have been studied for years by innovative scientists and engineers, modern photography is possible. The most dramatic change to the imaging chain in the past century has been the development of electronic imaging, which has altered the sensor, processing, and display elements of the chain. Historically, the sensor was light-sensitive film, processing required chemicals, and the display was on photographic paper. For most imaging chains today, the

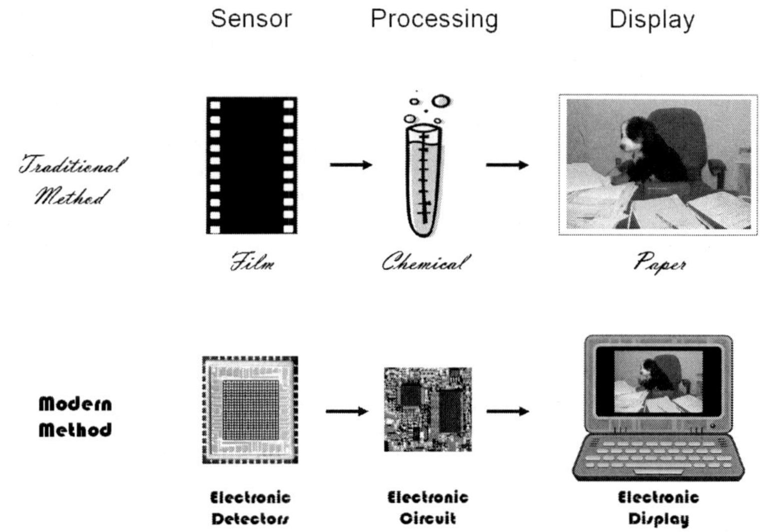

The sensor, processing, and display elements of imaging have changed in the past few decades from the traditional elements of the previous 100 years.

sensor comprises electronic detectors, processing uses software on electronic circuits in computers, and the display is on an electronic monitor. To help us better understand the creation of a photograph, let's take a brief look at the innovations that lead us to where we are today.

1.2 A Brief History of Photography—It All Started with a Hole in the Wall

The word *photography* is derived from the Greek words for light (photo) and writing (graphy) and was first introduced by Sir John F. W. Herschel in 1839 as he looked for ways to preserve an image once it was captured. But we're getting a little ahead of ourselves because first we need to describe a device that could create an image of a scene to be recorded. For that story we need to go back in time more than 2000 years.

Ancient philosophers, such as Mo Di (470–391 BCE) in China and Aristotle (384–322 BCE) in Greece, described the formation of an image when light is projected through a simple pinhole. The device that could create an image using a pinhole was called a camera obscura, from the Latin words for dark (obscura) and room (camera). The term *camera* used today is simply a shortened version of camera obscura. Ibn Al-Haitham (965–1040), known in the West as Alhazen, was the first person to describe how a camera obscura worked. He demonstrated that a small pinhole in a wall of a darkened room would project an inverted image on the opposite side of the room. This was entertaining for secretly observing people walking by the building, but it wasn't much more than a curiosity at the time. In 1490 Leonardo da Vinci documented a detailed description of the camera obscura, and by the 16th century European artists had found a practical use for the camera obscura by tracing or sketching the projected scene onto their canvas to help them paint more realistic portraits. Some artists considered this practice a form of cheating, but others saw it as a useful tool to help them sell more paintings.

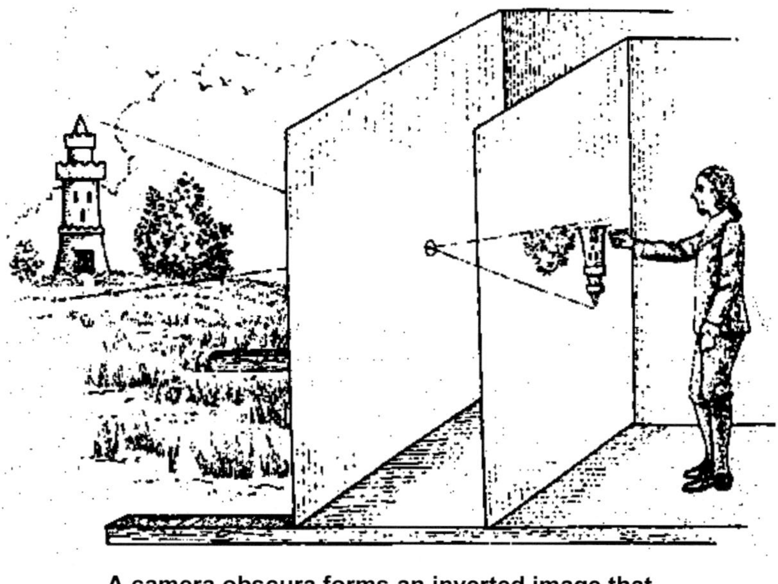

A camera obscura forms an inverted image that can be sketched using a tiny hole in the wall.

A camera obscura required the hole to be a very small pinhole to prevent the projected image from being blurred. The problem was that this small hole allowed only a limited amount of light to pass through, so the image was not very bright and needed to be viewed in a dark room. This problem was eventually addressed by using a glass lens with a larger hole in the camera obscura that redirected the light used to form an image; this also prevented the image from being blurred. *Lens* gets its name from the Latin word for lentil because its shape resembled that of the lentil bean. Around the same time that Ibn Al-Haitham was describing the camera obscura, glass quartz hemispheres were being used as reading stones by monks with poor vision to magnify text. As Venetian glass makers perfected the quality of lenses in the 14th century, reading stones were eventually placed in frames to make the first reading spectacles. The first use of a lens in a camera obscura is mentioned in 1550, when the Italian scientist Gerolomo Cardano suggested placing a lens similar to those made for spectacles into the hole of the camera obscura to improve the quality of the image. Shortly thereafter, the camera obscura was modified into a portable room that used a mirror to project the brighter image onto a drafting table, thus making it easier for the artist to sketch the scene. This transformation of the camera obscura from a large room to a portable imaging device was the first step toward cameras as we know them today.

Although advances were being made to the camera obscura, recording an image still required the skilled hand of an artist and included the artist's interpretation of the scene. An artist typically altered the truth, especially in portraits to make subjects more attractive, to the satisfaction of their paying customers. Recording a true image would require a material that was sensitive to light; it would also be necessary to "fix" the image after exposure, so that the material would stop changing with continued exposure to light. In 1724 Johann Heinrich Schulze discovered that silver salts, such as silver nitrate, would darken when exposed to light, but he had no way to stop the process to preserve the image.

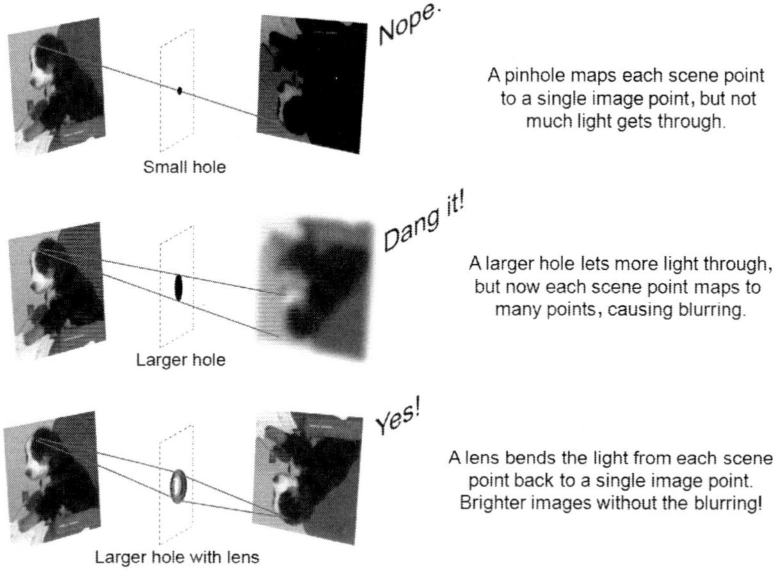

A lens was necessary to make brighter images without blurring.

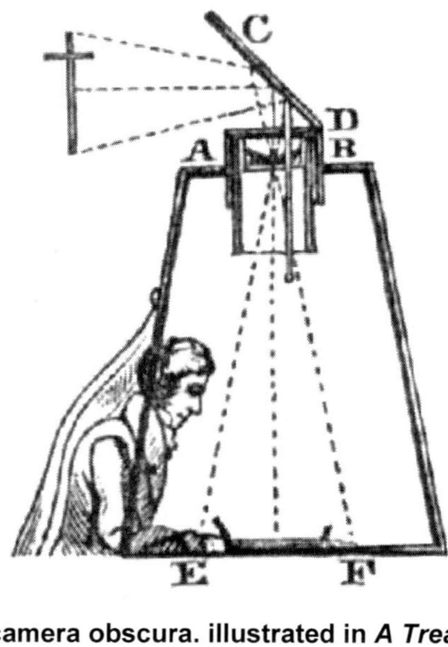

A portable camera obscura. illustrated in *A Treatise on Optics*, by Sir David Brewster in 1838.

The first major breakthrough for photography came in 1826 when French inventor Joseph Nicéphore Niépce created the first permanent photograph using bitumen, a tar-like substance, on a pewter plate. The bitumen would harden where it was exposed to light, and the remaining soft bitumen on the plate could be washed away with lavender oil to produce a negative image of the scene. The earliest surviving photograph from Niépce is entitled "View from the Window at Le Gras," imaged in 1826 at the commune Saint-Loup-de-Varennes using a camera obscura. The photograph is peculiar because both sides of the buildings appear to be illuminated by sunlight. This strange effect is the result of the long exposure time (between eight and twelve hours) required to produce a suitable image. The sun illuminated one wall in the morning and the opposing wall in the afternoon. Niépce knew that finding a person to sit still for more than eight hours was going to be a bit of a problem.

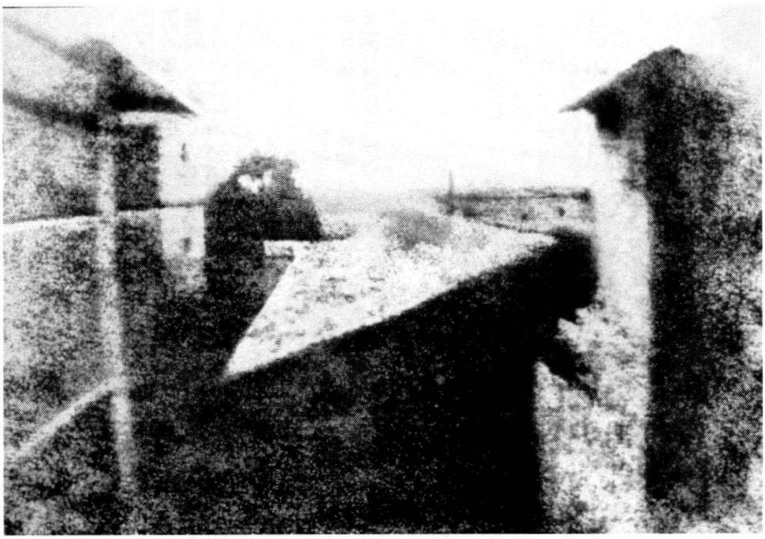

The earliest surviving photograph (1826). Not much to look at by today's standards, but this started it all!

In 1825 Niépce partnered with Louis Daguerre to develop a better process that did not require hours of exposure time. Unfortunately, Niépce died in 1833, but Daguerre continued the research. In 1837 Daguerre produced the first daguerreotype, a photograph created using light-sensitive silver iodide crystals on a copper plate. The exposure time needed to produce a photograph had been reduced to 15 minutes, which seems long by today's standards, but it was certainly better than eight hours! A year later he inadvertently captured the first picture of a person while photographing the Boulevard du Temple in Paris. The photograph required a ten-minute exposure time, so all of the moving people disappeared, but the man having his boots shined in the lower left of the picture stood still long enough to appear in the photograph.

Although the Daguerreotype process opened the door to modern photography, it still had many practical issues: exposure time was too long for portraits, photographs could not be copied, the process was not very portable, and the cost of equipment kept photography out of the hands of most people. There was also a health risk with the chemicals used, specifically the toxic mercury vapors used in the development process.

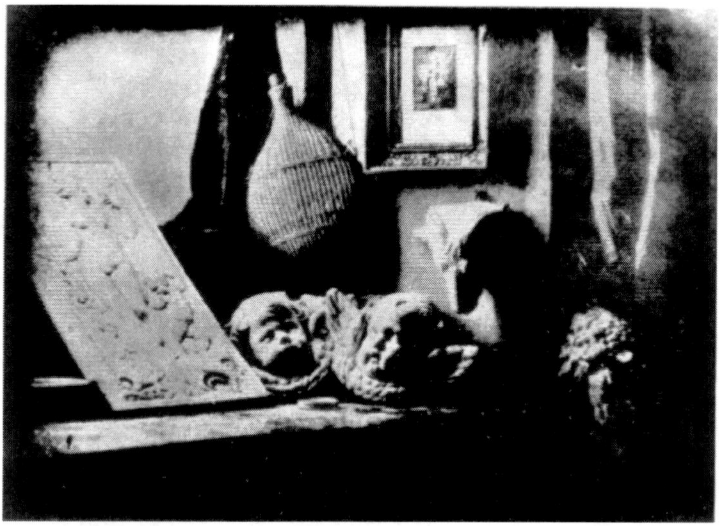

The first Daguerrotype (1837) made practical photography possible.

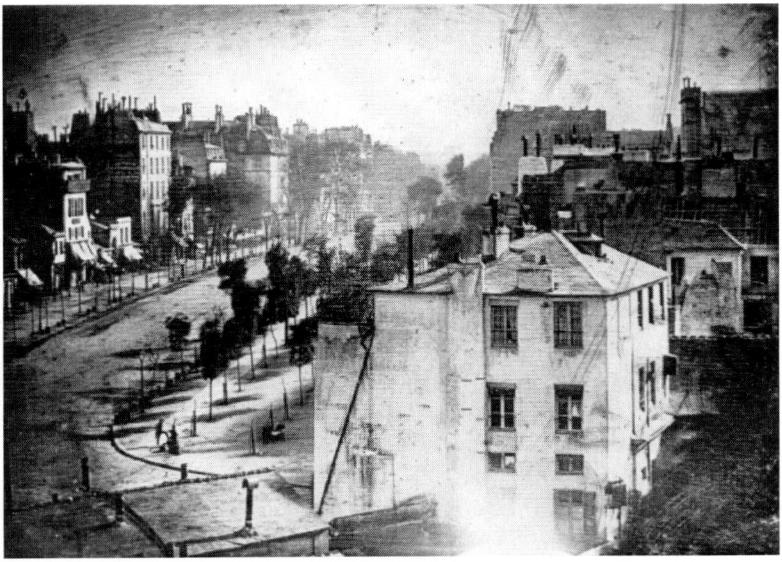

The first person to be recorded in a photograph was standing still to have his boots polished and is visible in the lower left of the picture (1838). He had no idea that he was making history!

The next advancement in photography was the development of wet plates, introduced by Frederick Scott Archer in 1851. A clean plate of glass was covered with a liquid light-sensitive emulsion, which only required a few seconds of exposure. Copies of the image could be readily produced from the negative image on the glass plate. The wet-plate process, sometimes called the collodion process, greatly simplified portrait photography. The key disadvantage, in addition to dealing with dripping plates in the camera, was that the plates had to be developed before the emulsion dried, which was usually less than ten minutes. This problem was solved with the invention of gelatin dry plates in 1871 by Richard Leach Maddox. During this time, cameras were becoming smaller and more portable, thanks to improvements being made to lenses and the sensitivity of emulsions.

Photography was finally available to the general public in 1888, when George Eastman introduced the Kodak camera, which used roll film instead of plates. Roll film was constructed using a thin and flexible plastic base for the dry light-sensitive emulsion that could be rolled without breaking. This film could be rolled tightly and placed in a small cartridge inside a camera. The Kodak camera could be held in the hand and had enough film for 100 photographs without switching in a new plate for every picture. To simplify the process even further, the photographer simply returned the camera with the film for processing after all of the pictures had been taken. As Eastman's advertising slogan stated, "You press the button, we do the rest." During the next 70 years, the quality of roll film improved, color roll film was introduced, and cameras became smaller and more automated. Unfortunately, people still wouldn't see the pictures they had taken until the film roll was completed and developed, usually days but sometimes months after the picture was taken.

In 1948 Edwin Land (founder of Polaroid Corporation) introduced the Land camera, an instant camera with self-developing film that could produce a photograph in minutes. While these advancements to the traditional chemical method of photography were being made, something else was happening that would revolutionize the future of photography.

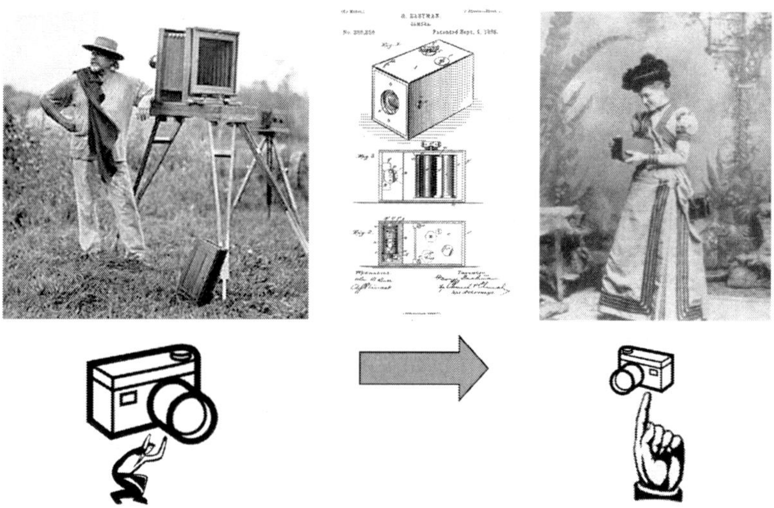

George Eastman's 1888 patent for the Kodak camera using roll film put photography in the hands of everyone.

A better way to record and preserve television shows soon became apparent as television became more popular in the late 1940s. At that time, shows were recorded by filming the television monitor on a device called a kinescope. In 1951 Charles Ginsburg at Ampex Corporation developed the first videotape recording (VTR) of a television show by recording the electronic signal from the television camera onto magnetic tape. Although not a digital image, the image was recorded electronically without the use of film. The first practical VTR with sufficient quality was the Ampex VRX-1000 (later renamed the Mark IV), and on November 30, 1956, it was used to produce the first tape-delayed television broadcast by rebroadcasting “Douglas Edwards and the News” to the Pacific Time Zone in the western United States.

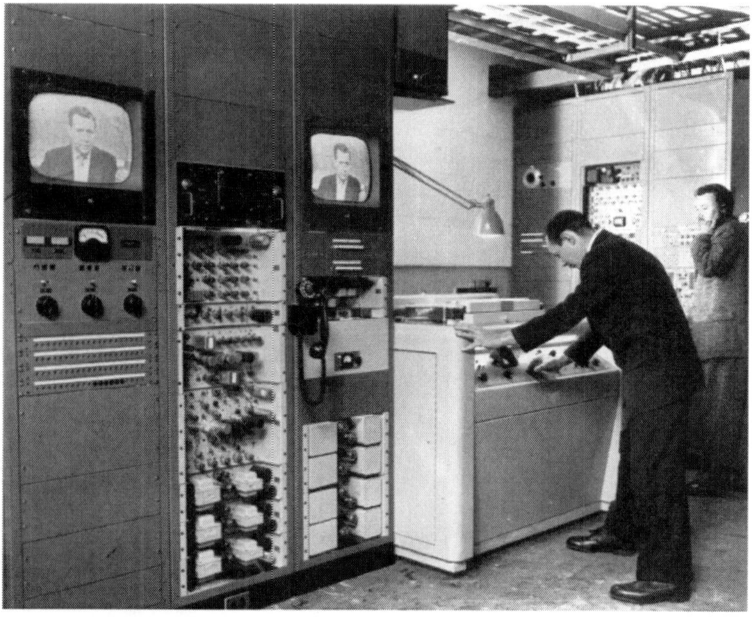

The first tape-delayed television show, “'Douglas Edwards and the News,” at CBS Television City in Hollywood, California (1956). The electronic signal from television cameras could now be saved and used to rebroadcast a show.

In 1957 Russell Kirsch created the first digital image in the world. Kirsch and his colleagues at the National Bureau of Standards (known today as the National Institute of Standards and Technology, or NIST) developed the first programmable computer, called the Standards Electronic Automatic Computer (SEAC), and created a rotating drum scanner that could read an image into the computer. A photograph was placed on the drum and scanned by a detector called a photomultiplier tube, which reacted to the light and dark areas of the photograph. The first photograph scanned was an image of Kirsch's three-month-old son Walden. The scanner divided the photograph into 30,976 regions (176 × 176) called picture elements, or pixels, and recorded a "1" if the region was bright or a "0" if the region was dark. These numbers were then stored in the computer and could be decoded to create a copy of the photograph.

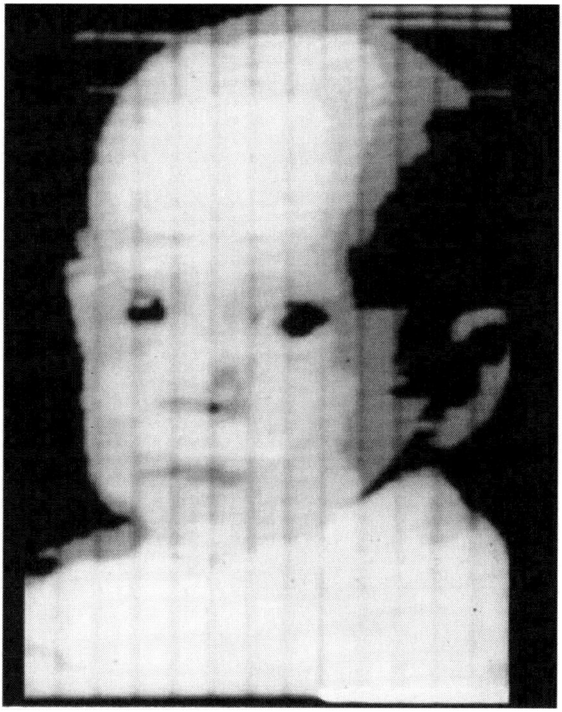

The first digital image was created by scanning a photograph in 1957. (Image courtesy of NIST.)

Working at Bell Labs in 1969, George Smith and Willard Boyle invented the charge-coupled device (CCD), a solid-state sensor that stores an electric charge proportional to the exposure to light. Although the CCD was originally designed as a new memory device for computers, its potential for imaging was soon realized. In 1975 Steve Sasson at the Eastman Kodak Company used a CCD to create the world's first digital camera to take filmless photographs. The camera took 23 seconds to capture its first image, which comprised 10,000 pixels on a cassette tape. A microprocessor converted the digital image into a video signal that could be played back and displayed on a television set.

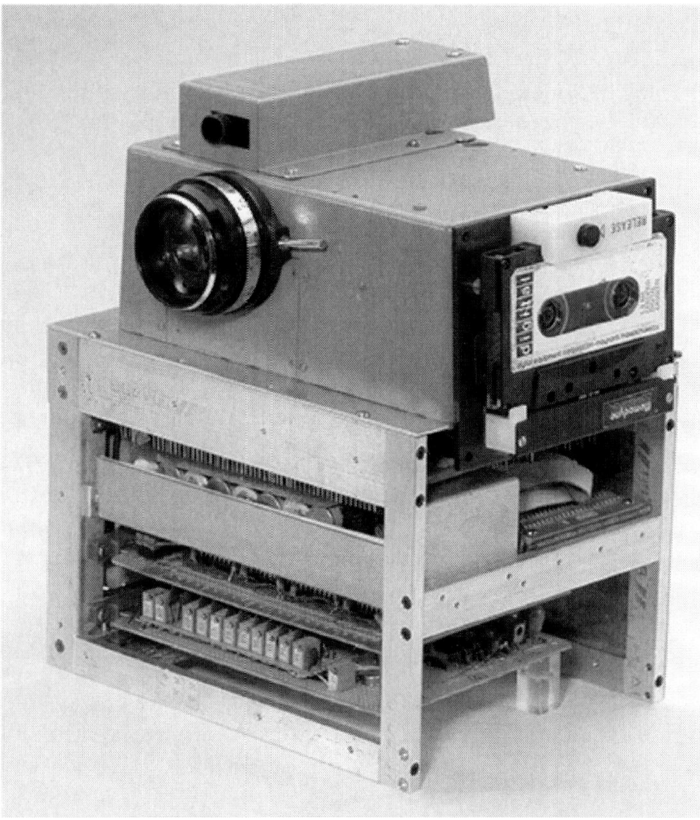

The first digital camera, created in 1975, was not quite ready to stuff into your shirt pocket.
(Image courtesy Eastman Kodak Company.)

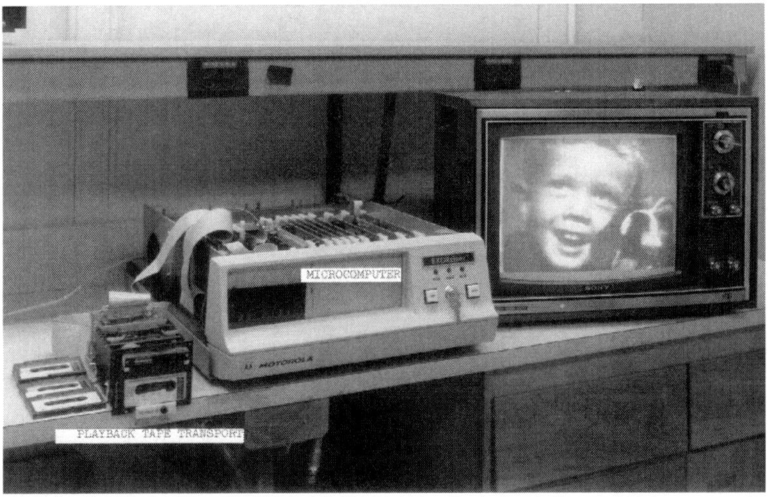

Display of an image recorded on a cassette tape from the first digital camera. (Image courtesy Eastman Kodak Company.)

In 1994 Apple introduced the QuickTake 100 camera (built by Kodak), the first color digital camera targeted for the general consumer that worked on home computers. The camera was only able to store eight pictures at full resolution, which was 640 × 480 pixels (about one third of a megapixel). There was no way to preview pictures, but the camera gave the general public their first experience of taking pictures and looking at them quickly on their home computer instead of waiting for film to be processed. Other companies quickly followed, and before long the image quality of digital photographs approached that of film. With the Internet revolution of the 1990s accelerating the need for digital images, digital cameras became less expensive and produced better picture quality. By 2003 digital cameras were outselling film cameras, and a mere two years later digital cameras were outselling film cameras by three to one.

Digital cameras have revolutionized the way images are communicated today, especially with their integration into mobile phones. When a newsworthy event takes place anywhere in the world, odds are that somebody is there taking photographs with a digital camera, and people all around the world will see those images within minutes. Satellites with digital cameras onboard such as the GeoEye-1 and WorldView-2 satellites, send back impressive images captured more than 400 hundred miles above Earth. These satellites snap pictures while traveling at speeds of four miles per second *without blurring them*!

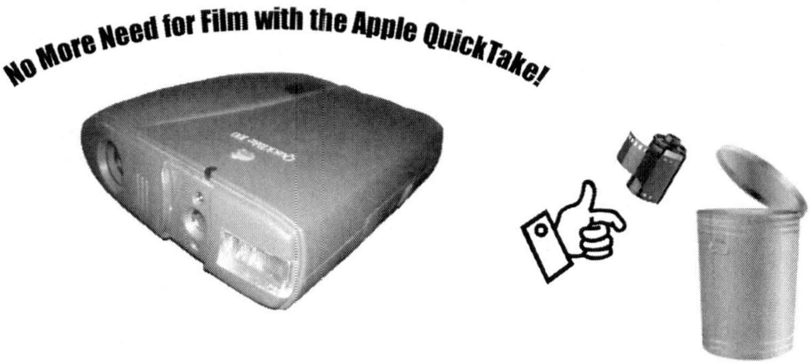

The Apple QuickTake 100 brought digital photography to the general public.

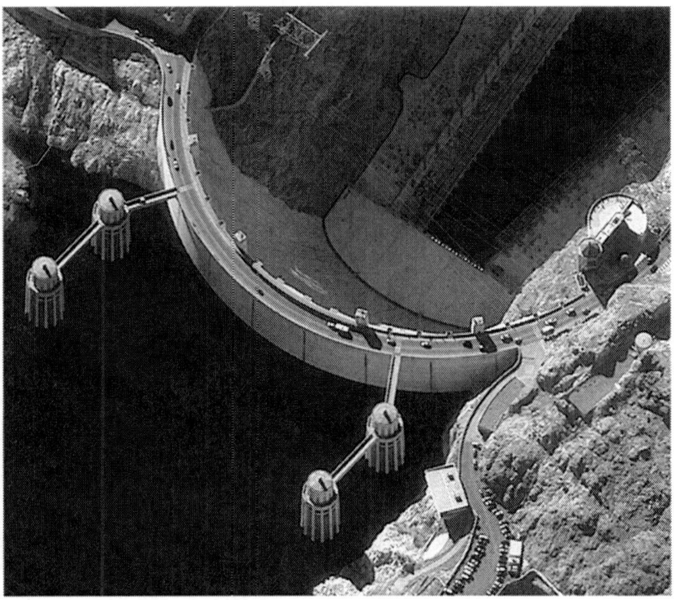

GeoEye-1 satellite image of Hoover Dam.
(Image courtesy of GeoEye.)

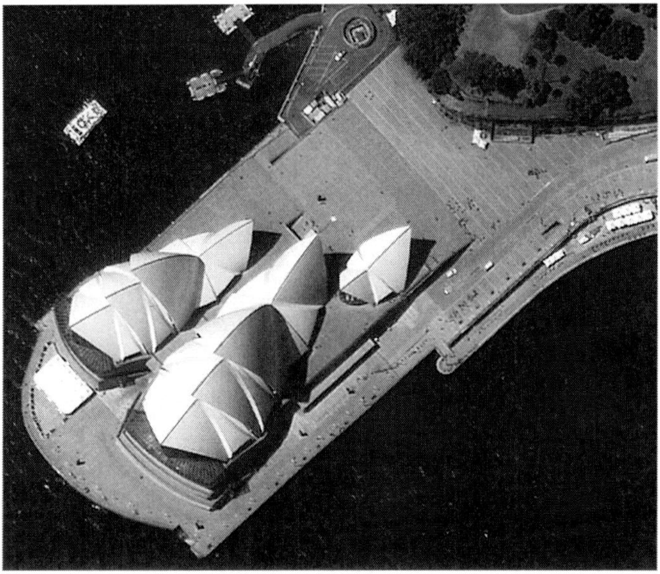

WorldView-2 satellite image of the Sydney Opera House.
(Image courtesy of DigitalGlobe.)

Chapter 2

Digital Images: Pixels and Bits

2.1 So, What Exactly Is a Digital Image Anyway?

The best place to start is to look closely at a digital image on a computer display. If you look at a black-and-white computer display through a magnifying glass, you can see that the image is made up of tiny dots called picture elements or pixels. Each pixel is a cluster of three dots (red, green, and blue) if you look at a color computer display. When you back away from the image, the individual pixels vanish because your vision blends the neighboring pixels together to form the image that you see.

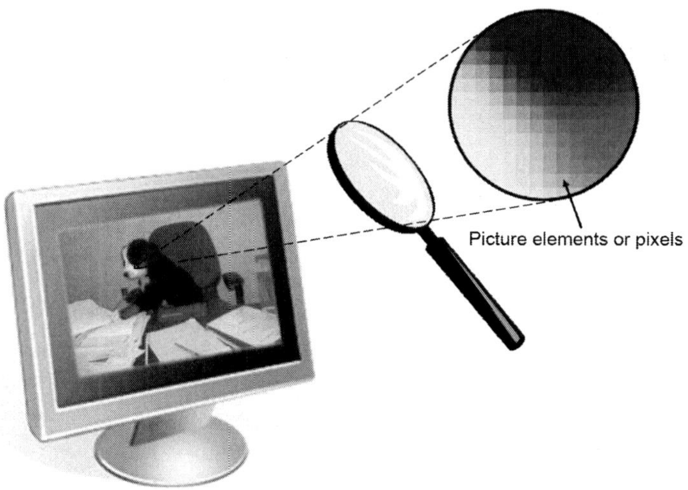

A digital image comprises dots called pixels that appear as different shades of gray.

Early computer games in the late 1970s used processors that were unable to handle many pixels, so the scenes generated were very blocky or pixilated. As the speed and memory of computers improved, the graphics that could be generated improved dramatically. It was a big deal in 1987 when Video Graphics Array (VGA) hardware allowed more than 300,000 (640 × 480) pixels to be displayed. Today we look back at early computer games and find it humorous that we were amazed at the limited graphics, but then again, that was only the beginning!

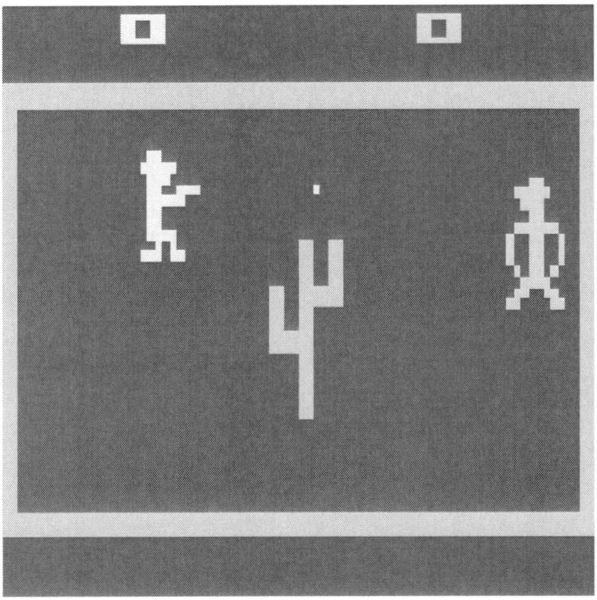

Early computer games, such as this Atari game, were limited to a few thousand pixels, so objects appeared blocky. The graphics seem laughable compared to today's games, but at the time we were excited to have it this good!

The size of a digital image is usually stated in terms of the number of pixels that make up the image. A one-megapixel digital camera can capture photographs using one million pixels, whereas a ten-megapixel digital camera can capture photographs using ten million pixels. More pixels are needed to see more detail in an image, but additional pixels can also be used to see more of a scene.

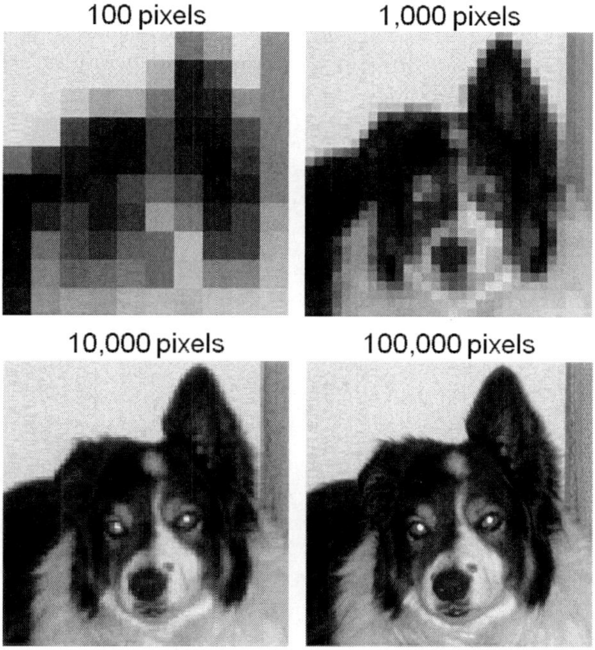

More pixels are necessary to see more detail in the image.

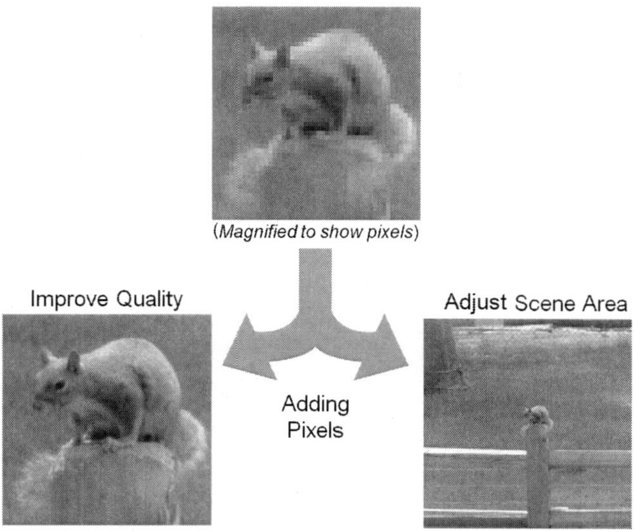

More pixels can be used to add more detail to a photograph, or they can be used to capture a bigger scene.

A digital image on a computer display then is just a bunch of lit-up dots (pixels) that vary in brightness to create an image we recognize. How does the computer know how bright to make each pixel on the display? To a computer, a digital image is actually just an array of numbers called gray levels. Each number tells the computer how bright to make each pixel on the display. Most computers use a range of numbers from zero (for black) to 255 (for white), giving 256 distinct brightness levels that can be used to display an image. To understand why 256 gray levels are used instead of, say 100, we need to talk about bits.

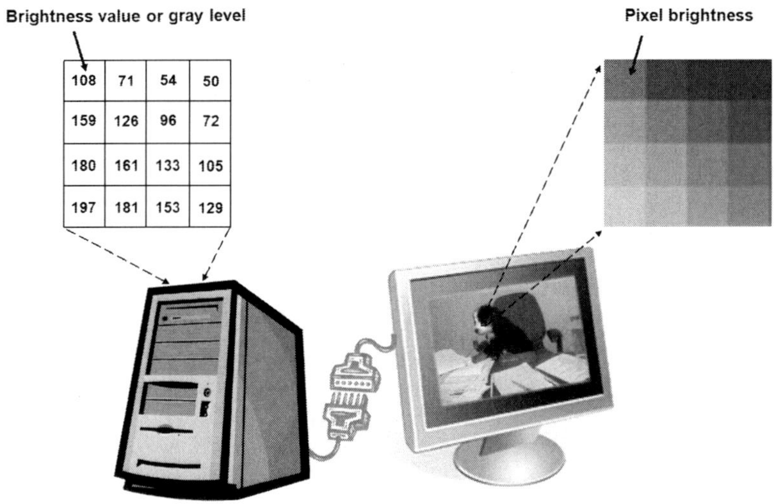

The numbers that make up a digital image tell the computer how bright to make each pixel when displayed.

At the most basic level, all a computer really knows is whether an electric switch is turned on or off, so the computer communicates in a language called binary. Binary means that the language is built on having only two ways to communicate information, such as yes and no, on and off, or 0 and 1. The gray level of a pixel must be communicated in binary digits, or bits, where each bit tells the computer to turn on or off a switch. The number of bits used to assign the gray level a numerical value is, therefore, equal to the number of switches that can be turned on or off. If only one bit is used, the digital image can only have two possible brightness levels: black when the switch is off, or white when the switch is on. If you use two bits, the first switch can be on or off and the second bit can be on or off, resulting in four possible brightness levels. Note that the number of possible gray levels increases by a factor of two for each additional bit used.

	Possible settings for one bit (one switch)	Binary number	Gray-level value	Brightness
1	off	0	0	Black
2	on	1	1	White

An image that uses one bit to quantify the gray levels of pixels can only have two brightness levels: either black or white, just as a light bulb is either dark or lit.

	Possible settings for two bits (two switches)		Binary number	Gray-level value	Brightness
1	off	off	0 0	0	Black
2	off	on	0 1	1	Dark Gray
3	on	off	1 0	2	Light Gray
4	on	on	1 1	3	White

An image that uses two bits to quantify the gray levels of pixels can have four brightness levels.

Using more bits allows for more shades of gray, but increases the file size of an image, which requires more disk space to store the image and more time to send or receive the image. The following figure shows the effect of using more bits to provide more gray levels in an image. Note the contours that appear on the wall when there are not enough gray levels to allow for a smooth transition between dark and bright regions. The 256 gray levels provided by 8 bits are generally sufficient for good image quality, so this has become the standard used today with most digital cameras. Computers also like to store information in units of 8 bits, called a byte, so using one byte to span gray levels is convenient.

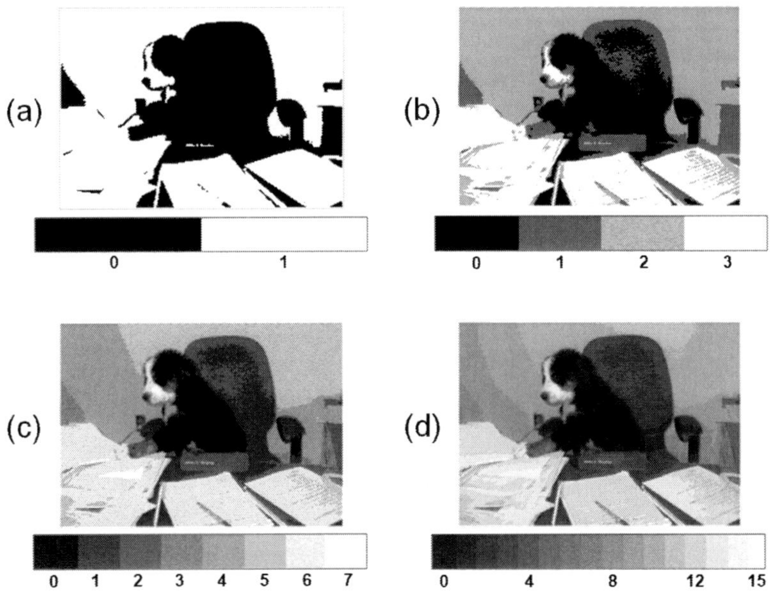

An image using (a) 1 bit, (b) 2 bits, (c) 3 bits, or (d) 4 bits to span gray levels from black to white.

(*Figure continued on next page.*)

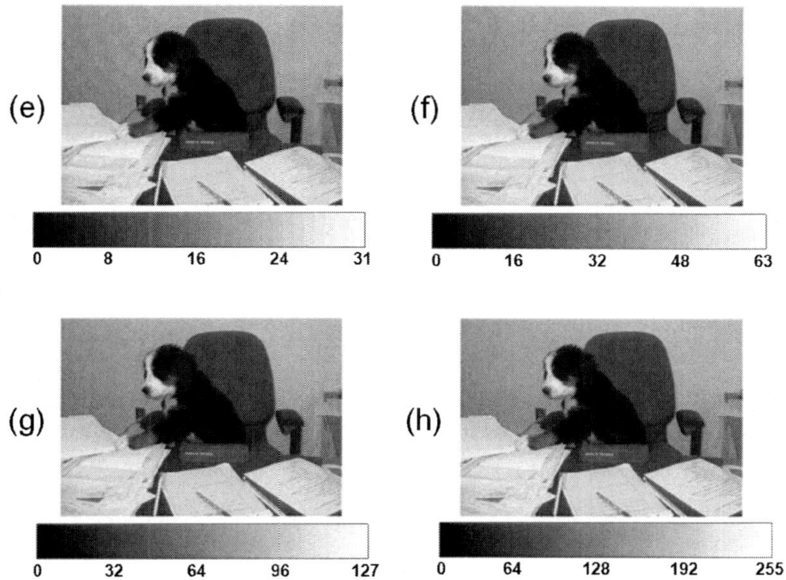

(*Figure continued from previous page.*)

An image using (e) 5 bits, (f) 6 bits, (g) 7 bits, or (h) 8 bits to span gray levels from black to white.

The file size of a digital image is almost always expressed in bytes, such as kilobytes (one thousand bytes) or megabytes (one million bytes). The number of bytes used to store an image depends on the number of pixels in the image and the number of bits used to span gray levels. It also depends on the amount that the image was compressed, a way of creating smaller file sizes by removing nonessential information. Cameras built for scientific measurements or for imaging Earth from space generally need more than 256 gray levels because the difference between bright and dark objects can be quite extreme, and there needs to be enough granularity between the different gray levels to record subtle differences in very dark and very bright objects within the same image.

Color images are created by combining three arrays of numbers, with one of the three arrays representing the red part of the pixel, another representing the green part of the pixel, and the third representing the blue part of the pixel. Why red, green, and blue? These three colors make up the three primary colors, and combinations of these three colors can make all of the colors in the rainbow. For example, equal parts of red and green with no blue will make yellow. When all three color values are the same, a shade of gray is created; black is produced when all three colors are zero, and bright white is created when all three colors are 255 (assuming 255 is our highest gray level). For simplicity, we will focus on black-and-white digital images, but keep in mind that a color image requires a combination of three images in the computer to create the colors.

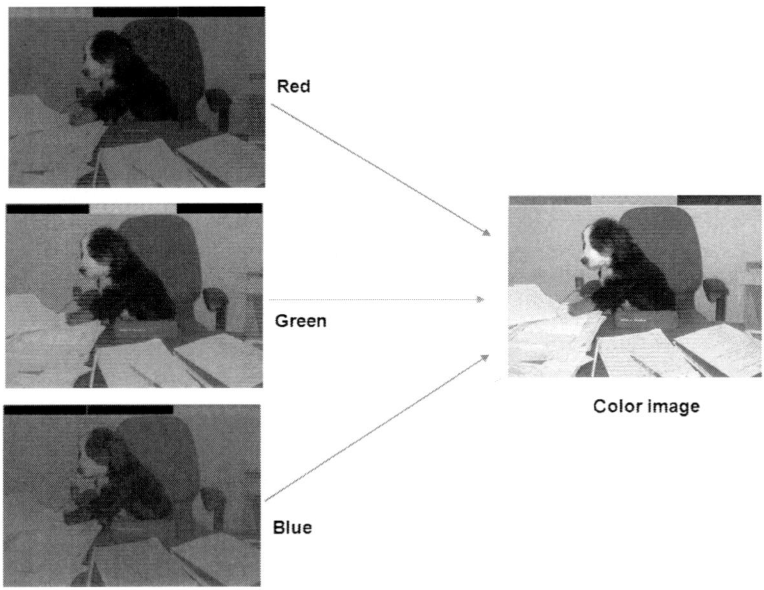

A color image is actually a composite of red, green, and blue images.

A digital image to a computer is just an array of numbers that range between zero and 255, where each number tells the computer how bright to make each pixel on the display. But now comes the hard part, specifically understanding how these numbers are created.

2.2 Making the Numbers—How to Make a Digital Image

The quickest and easiest way to create the numbers that make up a digital image is, of course, to let the camera do the work for you. The digital camera converts light from a scene into an array of numbers that, when displayed, looks a lot like the scene we saw with our own eyes. This process is discussed in detail in subsequent chapters, since that is the subject of this book, but what if you can't get the picture you want with a camera? For example, if someone wants a picture of Godzilla stomping out a cigarette, can they just make up the numbers on their own to make that picture? You bet they can! But it's more difficult than it sounds, and without any thought given to the numbers, the image created would not be very interesting. Let's start by just throwing numbers (remember they need to be between zero and 255) into an array and displaying it in the following figure. It might be interesting to abstract artists, but it certainly doesn't look like anyone I know.

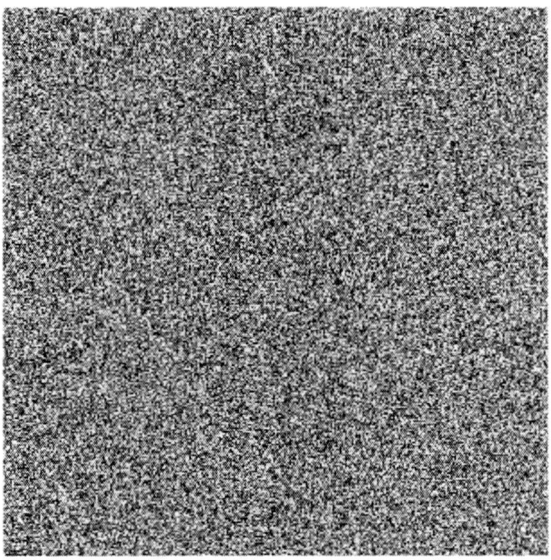

Just making up numbers to create a digital image results in an uninteresting picture (to most people, anyway).

To form a meaningful picture, the pixels must have gray-level values that collectively form a recognizable object, similar to creating a mosaic, where each tile is a picture element of a particular shade of gray necessary to form the desired picture. A paint-by-numbers canvas that guides us in placing gray-level values at each pixel location would be ideal. (Remember, for color pictures three numbers are needed for each pixel.) However, choosing the number to put in each pixel to give us the desired brightness is going to take a long time, unless we have a limited number of pixels or gray levels to choose from. To put it in perspective, if we take ten seconds of thought to determine the necessary brightness for each pixel to create a 4-megapixel black-and-white image, and we worked 40 hours per week, it would take more than five years to complete the image. A color image would be more difficult, because not only are there three images to create (red, green, and blue), but now all of the possible color choices for each pixel must be considered. Each of the three images can have 256 possible shades, and when combined, they will create almost 17 million possible colors to choose from!

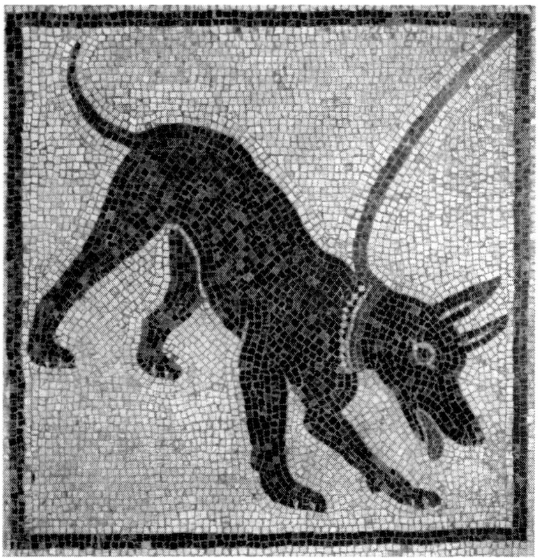

A digital image is similar to a mosaic, such as this one from Pompeii, because the tiles are essentially pixels of various gray levels.

Fortunately, computers come to the rescue with graphic illustration software to help put the numbers into an array much faster than we could on our own. When you use computer software to draw a computer-generated image (CGI), you are really just telling the computer what gray level you would like for each pixel. Drawing pictures free-hand on a computer is easy, but more sophisticated methods of creating an image are needed to add more realism and more complex image effects, such as reflections, shadowing, and texture.

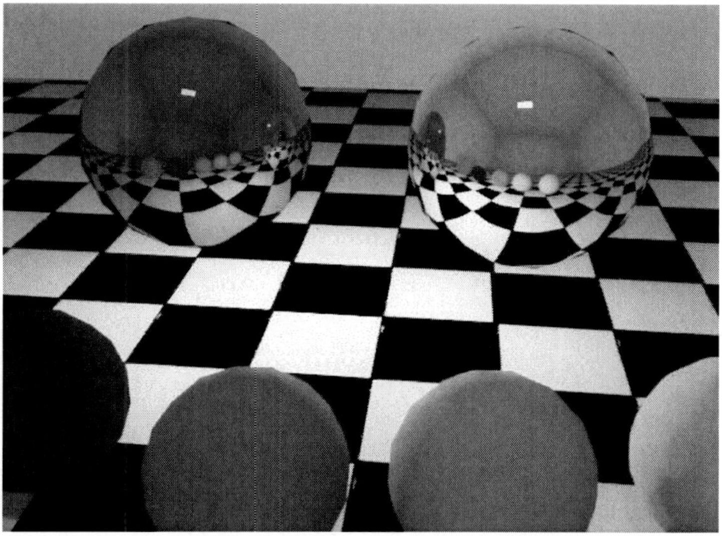

Sophisticated computer graphics software is needed to help get the gray levels for each pixel correct to create realistic reflections.

2.3 Is That Really a Picture of an Alien?

If the numbers in a digital image can be played with, it should be easy to make a fake image. We will define a fake image as either an image that is not created by a camera, or an image that originated from a camera but has been altered to change its meaning. If you cut and paste the head of a dog over the head of a person in a digital image, you are really just changing the gray levels of the pixels on the human head to the gray levels of the dog head. Software is readily available and easy to use, allowing us to have harmless fun altering images, but what happens when people alter images in an attempt to deceive us in an immoral way? Suddenly, it's not much fun anymore. As long as there is a quick buck to be made, people will continue to pass off fake images as real, especially when it comes to images of ghosts and extraterrestrial entities. Probably the most dangerous motive for generating fake images is to alter the public's perception of truth for political reasons.

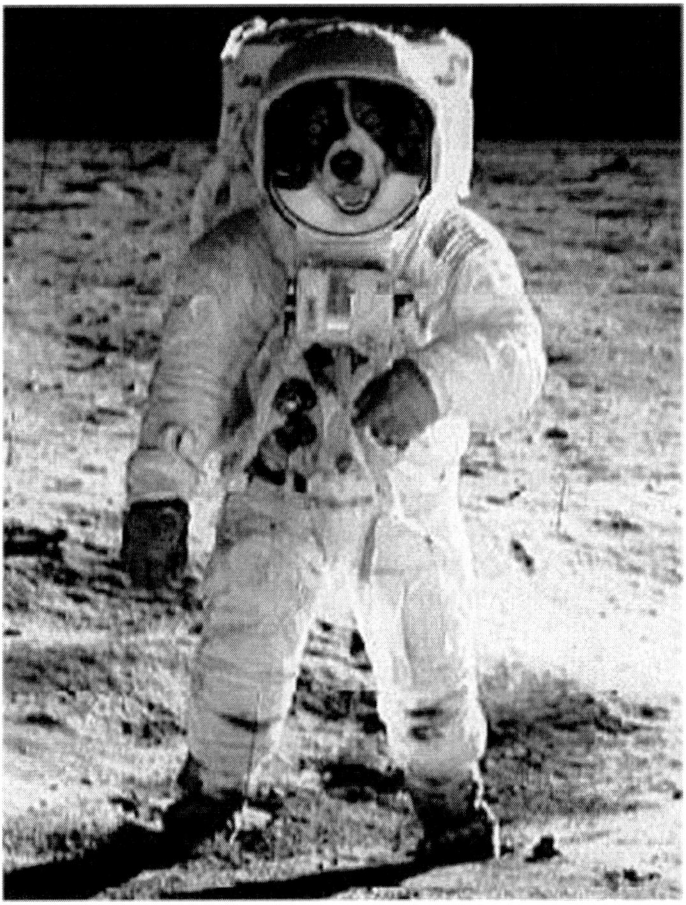

Creating fake images is usually quick, easy, and harmless fun.

Techniques used to fabricate a fake photograph have been perfected since the early days of photography. Early portrait photographers found that some clients, who were accustomed to having artists paint their likeness in a more favorable way, would not pay them for photographs that showed physical features they considered less than flattering. (How often have we looked at a photograph of ourselves and said, "Do I really look like *that*?") Photographers quickly learned that they would have happy paying customers if they applied darkroom tricks to simply touch up the photographs in a way that removed unflattering features. These techniques are still implemented today, with computers creating photographs of models and celebrities that are amazingly wrinkle free.

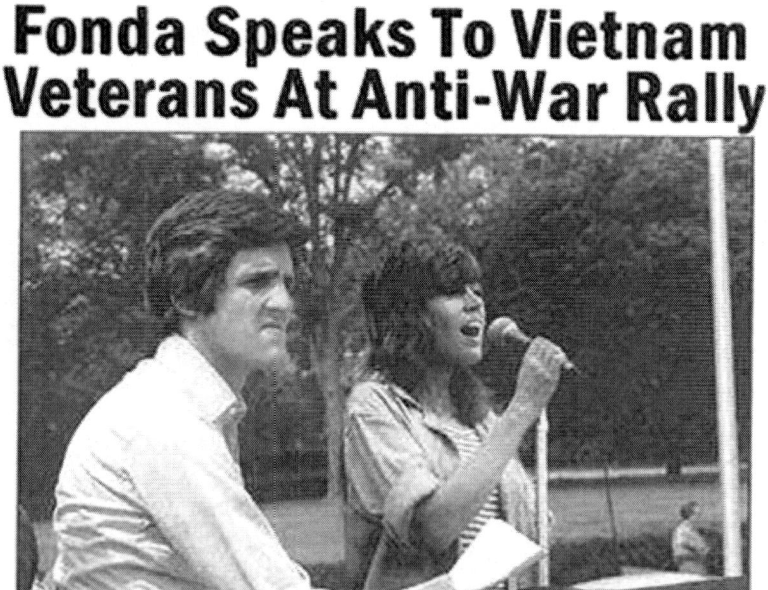

The fun ends when people create fake images to deceive us, such as this image created during the 2004 presidential election.
[Original image © Owen Franken/CORBIS (1972).]

So, how do we know when we're being duped? Unfortunately, it's becoming harder every day, because computer programs are getting more sophisticated at making up the numbers in a digital image that mimic a real photograph. However, we can catch most fake images if we understand how they are made, and if we have some knowledge of the characteristics that make them different from real images. A real photograph must be created by following the rules dictated by nature to create the image. If an image breaks the laws of physics, we know there is something suspicious going on.

Creating a computer-generated image and passing it off as real is very difficult. The real world is extremely complex, even to a computer, and accurately reproducing everything from the object detail to the camera imaging properties is very difficult. The devil is in the details, and having a computer figure out which gray-level numbers should be assigned to each pixel to create a detailed scene that looks realistic is extremely complicated.

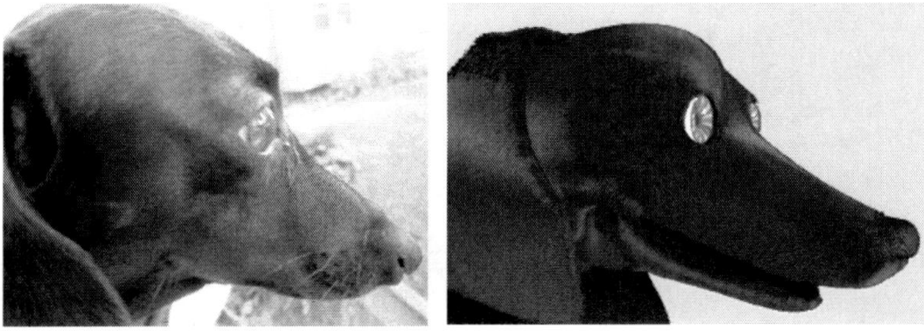

Photograph Computer-Generated Image

A computer-generated image generally lacks the realistic qualities seen in photographs captured with cameras.
(Courtesy of Katie Fiete.)

Most fake images are created by altering existing photographs, simply because it's fun and very easy to do on most computers. The unaltered portions of the image are real, so the overall image has a realistic look to it. Identifying an altered image is usually simple by noting the differences between altered and unaltered portions of the image. Remember, the entire image must be consistent with the science that created the image, and if the altered and unaltered parts have different properties, we know we have a fake on our hands.

Perhaps the most common inconsistency between altered and unaltered parts of an image is lighting and shadowing. The light source and the geometry between this light source, the object, and the camera, are probably different between the two photographs that were combined to form the fake image. For example, if the head of a person in a photograph taken with a flash camera is pasted over the head of an individual standing outdoors, the image will exhibit apparent differences in lighting and shadowing.

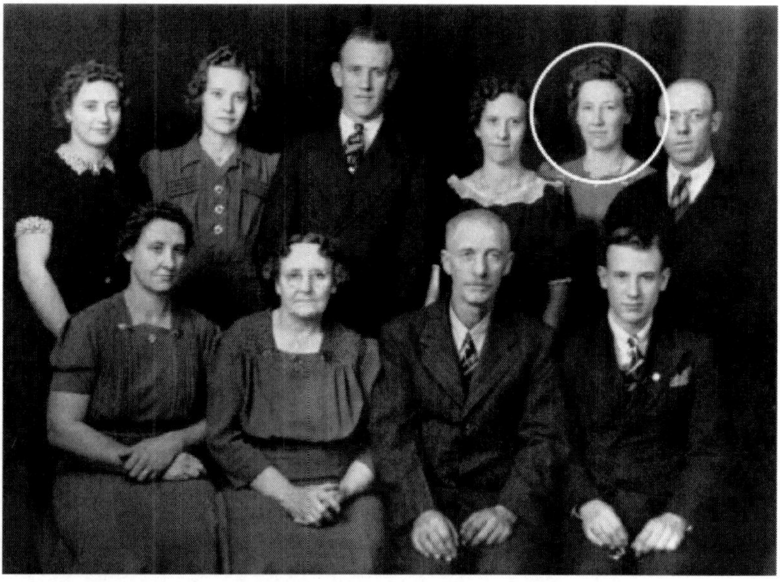

My grandmother (circled) was added to this photograph at a later date. Note that the flash should have illuminated her body but here it is inexplicably dark, revealing inconsistent lighting conditions in this photograph.

Another common error found in altered photographs is an inconsistent blurriness to the edges. A camera blurs an image by blending the brightness of neighboring pixels in a prescribed way that is often missing in altered images. If an object is cut and pasted onto another image, the relationship between neighboring pixels along the edge of the inserted object will be awry, and the edges will not be blurry enough. We know from our experience of viewing real photographs that this effect isn't normal, and the alteration will be obvious. To make the alteration less obvious, the edges around the inserted object can be smudged or blurred to simulate that blurring effect of the camera on a real edge in the scene. However, this blurring is usually inconsistent with blurring that the camera imposed on the rest of the photograph.

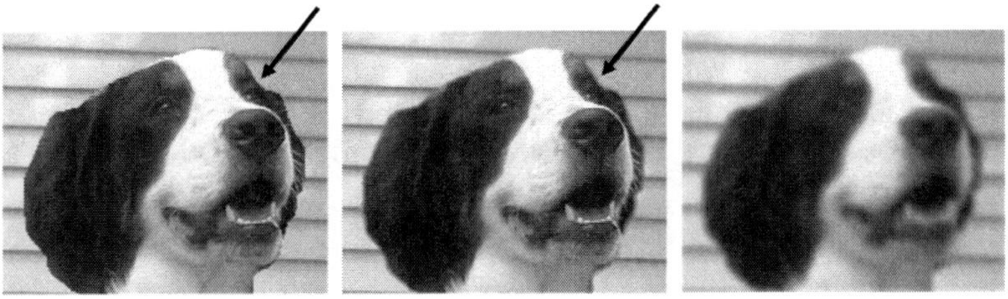

Edge of inserted object is too obvious.

Blurred edge is less obvious but not consistent with the rest of the image.

Entire image would need to be blurrier for blurred edge to be consistent.

Blurring the edges of an object inserted into an image can make the alteration less obvious, but it is still inconsistent with the amount of blurring created by the camera in the rest of the image.

The most difficult images to properly identify as fake are those that are real photographs with altered interpretations, such as a photograph of a Frisbee that is claimed to be an alien space ship. These photographs pass all of the scientific tests because they are, after all, real photographs. All of the gray levels in each pixel follow the laws of physics for creating an image. The challenge here is identifying inconsistencies between the photograph and the story that the owner is telling, which unfortunately will always prove to be difficult until better lie detectors are invented. In the end, the truth behind these fake photographs is not really known until the owner admits that the photograph is a fake.

When we see a picture of an alien, can we prove that it is real or fake? Unfortunately, we can never prove that an image is real, we can only prove that it is fake if it is not consistent with the characteristics of a real photograph. Identifying fake images becomes more challenging every day as computers become more sophisticated and modeling the imaging properties of cameras becomes more accurate. Remember, as long as the numbers that make up a digital image have all of the properties of a real-world photograph, maybe it is a real picture of an alien!

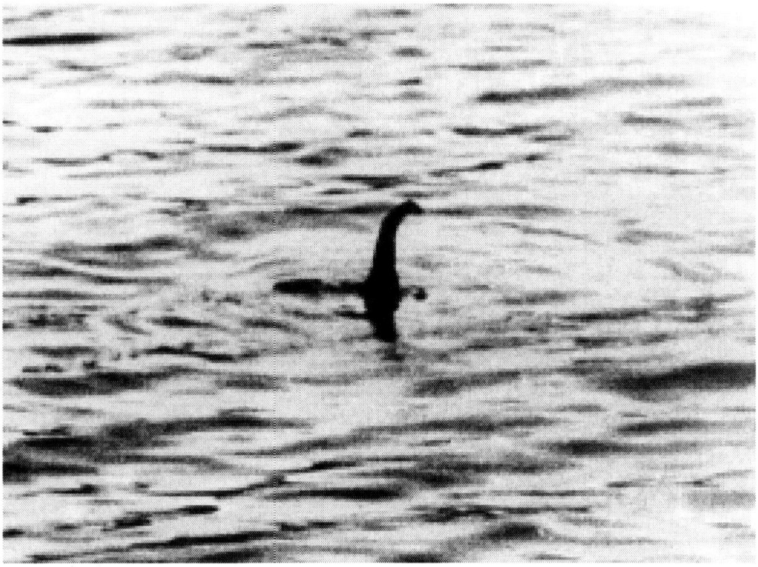

The truth behind the "Surgeon's Photograph" of the Loch Ness Monster, taken in 1934, was always in question until the hoax was revealed in 1994. The photograph was physically unaltered, so proving that it was fake required a confession from someone involved. (Photograph was claimed to have been taken by surgeon Robert K. Wilson.)

Chapter 3
Light: The Essential Ingredient

From experience we know that our pictures turn out well when a scene is brightly illuminated. Simply stated, light provides the energy necessary to record an image, so we need good lighting. Lousy lighting makes lousy pictures. Just as we find it hard to do good work when we feel low on energy, the camera has a hard time making a good picture without enough light energy. We need to make sure that the light comes together just right, or else the picture will still be lousy, even if the scene is sufficiently lit.

Another important consideration is the behavior of light. To design a camera that can form the best image possible, camera designers need to understand how light behaves as it propagates to the camera, through the camera, and finally onto the imaging sensor.

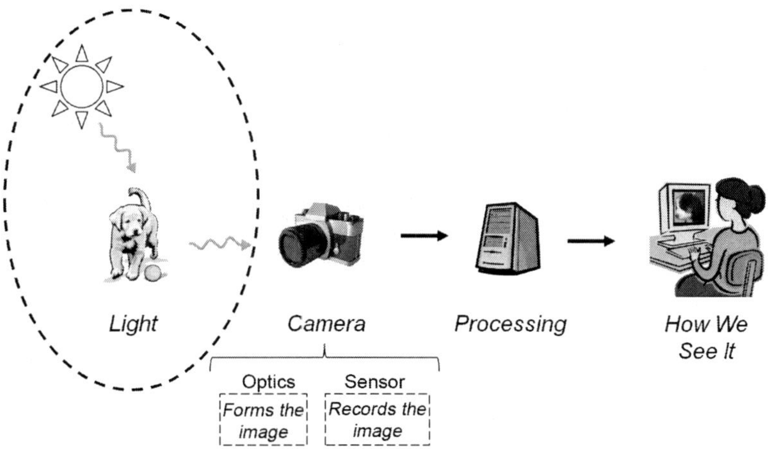

The imaging chain starts with seeing the light.

3.1 A Heavy Understanding of Light

We understand that light is the source of energy that creates the image, but what exactly is light? Dictionaries define light as something that makes vision possible. This is helpful when we think of our eye as a camera, but a definition that applies to all cameras in general would be better. Scientific dictionaries define light as electromagnetic radiation of any wavelength that travels in a vacuum with a speed of about 186,281 miles (300,000 km) per second. So, light moves pretty fast—fast enough to go 7.5 times around the earth in one second. But what is electromagnetic radiation, what is a wavelength, and how do we use them to make pictures?

Electromagnetic radiation is intertwined electric and magnetic fields that travel in the form of a wave. We describe a wave by the amplitude (the height of the wave) and the wavelength (the length from one peak to the next). Often it is more convenient to describe the frequency of the traveling wave (the number of up and down cycles that occur in one second, expressed in units of hertz) rather than the wavelength.

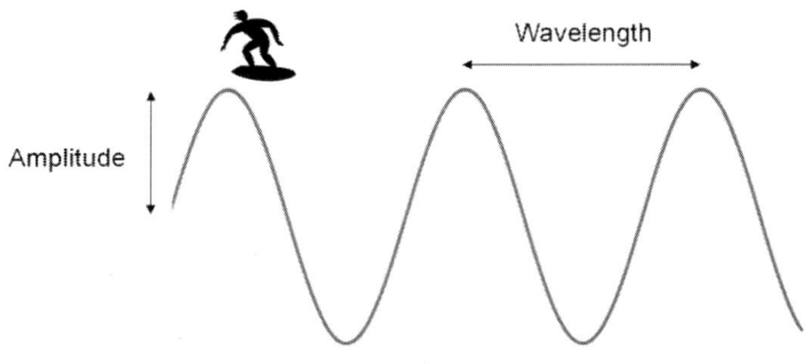

Waves are described by their amplitudes and wavelengths.

An electromagnetic wave is classified by its wavelength as a gamma ray, x ray, ultraviolet (UV), visible, infrared (IR), microwave, or radio wave. The range of electromagnetic waves that we can see with our eyes is called the visible spectrum and falls between the UV and IR spectra. The visible spectrum is also the dividing line between electromagnetic radiation that is generally harmful and that which is generally harmless to us. Shorter wavelengths contain much more energy and thus are more dangerous to humans than are longer wavelengths. Most of us are very aware that gamma rays are the most energetic, thanks to science fiction writers and the Hulk. Sunburns are actually a form of radiation damage to our skin from the ultraviolet light of the sun.

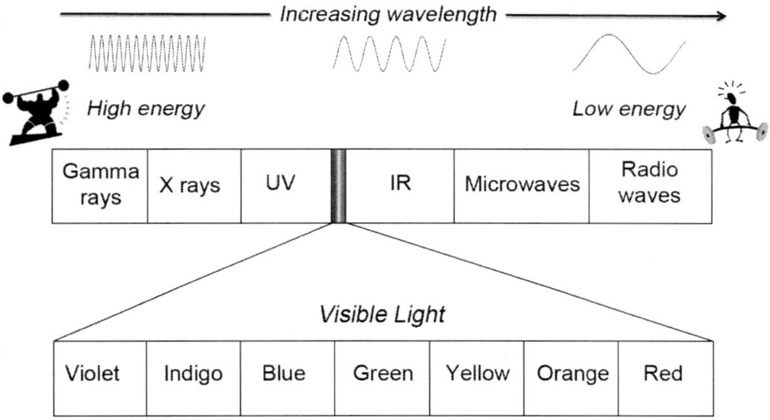

From gamma rays to radio waves, they are all part of the electromagnetic spectrum.

The visible spectrum spans the colors of the rainbow, with the longest wavelengths appearing as red and the shortest as violet. As a child we were told to remember the name of our friend Roy G. Biv, whose letters spell out all of the colors in the rainbow—red, orange, yellow, green, blue, indigo, and violet. Sir Isaac Newton published experiments in 1672 to show that sunlight contains all of the colors in the rainbow, and that a glass prism can separate the colors from each other. Before Newton's experiments, it was believed that *white light* from the sun was colorless and it was the prism that added colors to the light. Water droplets in the air act as prisms that separate the colors in sunlight and cause us to see each color in a different location. Light separated into a rainbow extends beyond the colors that we see, but the wavelengths that fall outside the visible spectrum remain invisible to us.

We will focus our attention on digital cameras designed to form images using visible light. It should be noted that all electromagnetic waves, regardless of their wavelength, travel at the speed of light.

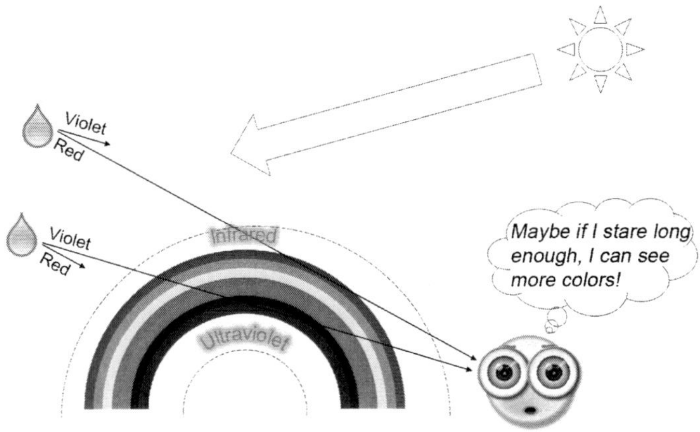

Water droplets spread out the different wavelengths of light from the sun, but our eyes can only see those that are in the visible spectrum.

The color composition of light is usually expressed in terms of monochromatic or polychromatic. The word monochromatic comes from combining the Greek words *monos* (solitary) and *chroma* (color), so monochromatic light has only one wavelength or color of light. The word polychromatic comes from combining the Greek words *poly* (many) and *chroma*, so polychromatic light is composed of many wavelengths or colors of light. Light emitted from lasers is monochromatic, whereas light emitted from the sun is polychromatic. Another term widely used is panchromatic, a word that combines the Greek word *pan* (all) with *chroma*, so panchromatic light contains the range of wavelengths for all colors across the visible spectrum.

The sun is the main source of light for outdoor imaging in the day time. You may hear scientists estimating the sun as a "blackbody" to help simplify the calculation of the light energy coming into the camera, but this seems like an oxymoron because the sun is anything but black. A blackbody is a theoretical body that absorbs all of the light that falls on its surface; hence, it would appear black if you shined a light on it. However, the blackbody also radiates light energy in a continuous spectrum based on its temperature. If it radiates light in the visible spectrum, we can see it with our eyes. The German physicist Max Planck worked out an equation that lets us calculate the energy and colors from a blackbody. Although there are no true blackbodies in the world, we are fortunate that the light from the sun fits the blackbody model pretty well. It's no coincidence that the light sensitivity of the human eye is centered very near the peak wavelength emitted for a blackbody at the sun's temperature (about 10,500°F or 5,800°C). Our eyes are a great match for a world illuminated by light from the sun!

When we see waves rippling in a lake, we know that the water is being disturbed in an oscillating up and down motion as the waves travel. When we hear the sound of a whistle, we know that the air has been disturbed in an oscillating compression and decompression motion as the sound moves. But what is disturbed when an oscillating light wave moves? Well, that's the funny thing—the answer is nothing. Scientists in the 19th century thought that all space must be comprised of a substance they called ether, and so light waves must be disturbing the ether. A famous experiment conducted by Albert Michelson and Edward Morley in 1887 at what is now Case Western Reserve University in Cleveland, Ohio, proved that this ether did not exist. We just have to accept that light is a self-propagating wave, and that this is one of the strange properties of light.

Another strange property of light is called wave–particle duality. Some people like to think of light as a particle of energy, called a photon, which has the properties of a wave, while other people like to think of light as a propagating wave of energy, having the properties of a particle. To make it simple from a camera's perspective, we can think of light as a wave while the camera collects it, but will think of it as a particle when we need to convert the collected light energy into a digital image.

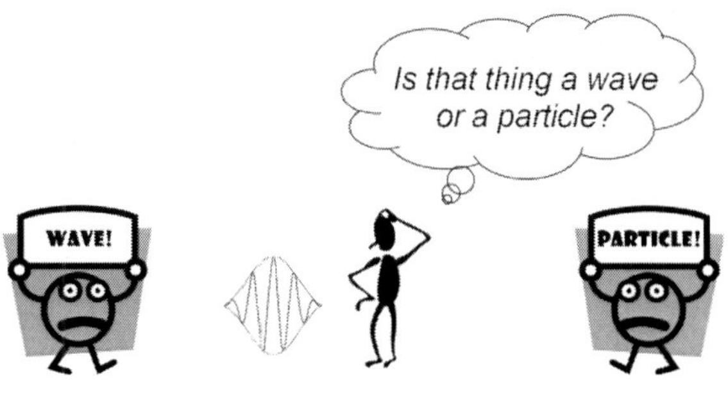

Light behaves like both a wave and a particle at the same time. It's probably better not to think about it too much.

3.2 Light Interrupted

Light travels happily through space until it encounters matter that it must interact with, similar to bumping into that strange relative that you were hoping to avoid at a family reunion. When light runs into a material, it can bounce off the surface (reflection), pass through the material (transmission), or become consumed by the material (absorption). Reflection and transmission can also be accompanied by scattering, which deflects the light in multiple directions. Transmitted light that is partially blocked by an edge spreads out behind the edge in a process called diffraction. Any combination of these effects is possible, so keeping track of light quickly becomes complicated. For the purpose of creating a photograph, reflection and transmission are important for redirecting light to form an image, while absorption is important for capturing light energy to record an image.

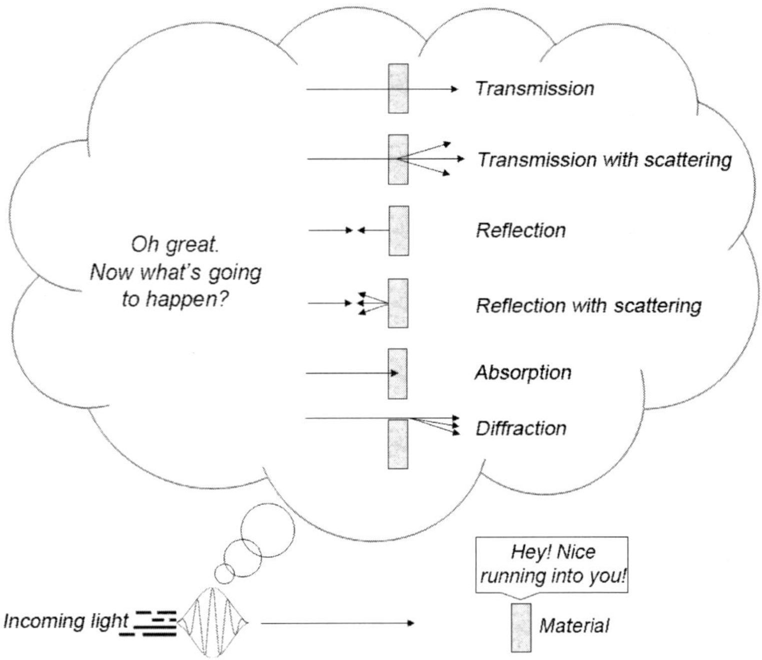

Light travels happily along until it runs into a material that it must interact with.

Almost everything that we see in our natural surroundings is the result of light reflecting off of a surface. An object that reflects most of the light back into the camera will be brighter in the image than an object that reflects only a small proportion of the light. A very nice and useful property of reflection is that a reflective surface redirects light away from its surface in a direction that has the same angle from which it came. This is called the law of reflection, which describes the property that is also referred to as specular reflection. The law of reflection makes it easy to determine where light is going when it reflects off of a surface, assuming we know the tilt of the surface, which we might not know at a microscopic level.

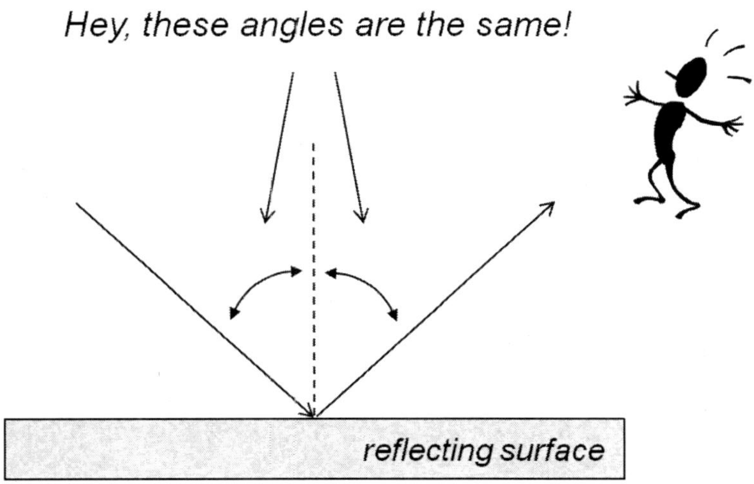

Light reflects away from a surface at the same angle that it came in on.

Most surfaces have a roughness to them comprised of tiny tilts; these tilts cause diffuse reflection by scattering light in different directions. This is actually more pleasing to our eyes, because we do not want to see every surface reflecting light like a mirror. But if we do want light to reflect from a mirror, it's important for the mirror to have a very smooth surface, so that the light will go precisely where we want it to go. This is very important when mirrors are used to form images, such as those used in large astronomical telescopes.

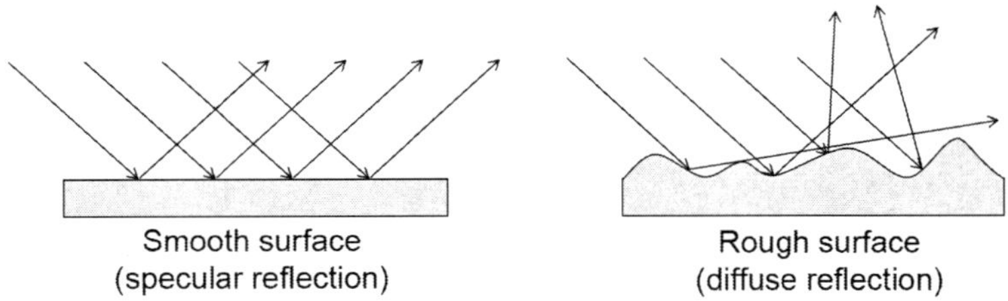

Specular reflection from a smooth surface causes objects to look shiny, while diffuse reflection from a rough surface scatters light and dulls the sheen.

When light passes from one transparent material to another, it bends and changes direction. This bending is called refraction, and the amount of bending is determined by the index of refraction, which is a measure of how much the material slows down light as it travels from a vacuum into the material. The index of refraction n is the ratio of the speed of light in a vacuum to the speed of light in a material. For example, water slows down light to three-quarters of its speed in a vacuum, so the index of refraction for water is 4/3. We can determine how much the path of light will be altered when light enters a material using the law of refraction, also known as Snell's law. The amount of bending from refraction is also dependent on the color of the light, which is good news for those of us who enjoy rainbows. It is this wavelength dependence on the amount of refraction, called dispersion, that causes each color in sunlight to bend differently, thus separating the colors into the beautiful rainbow that we see.

When light transmits through a medium, the scattering of different colors that can occur also must be considered. In general, scattered light is not favorable because the light goes in uncontrolled directions. Scattering of transmitted light is the answer to the age-old question, "Why is the sky blue?" As light from the sun passes through our atmosphere, Rayleigh scattering causes blue light to scatter and bounce around high in the sky like a pinball, while the other colors pass through the atmosphere. When looking up at the sky, we see the blue light that is bouncing around up there before it finds a way to our eye. Light is still scattered in the atmosphere at night, but without the flood of blue light from the sun scattering in the sky, there isn't enough light to hide the stars. It's interesting to note that if blue light did not scatter around in the sky, the daytime sky would appear black, and the stars would not be hidden by the blue light.

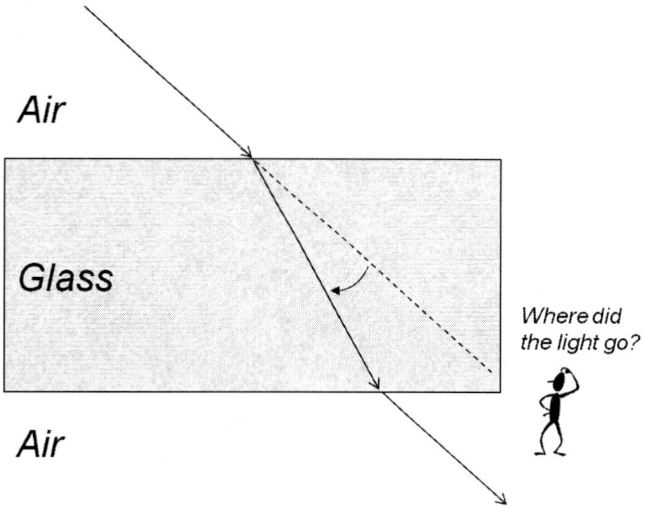

The property of refraction alters the path of light through a transparent material, such as glass or water.

3.3 A Horse of a Different Color

It's important to note that the reflection, transmission, absorption, and scattering of light can be different for different wavelengths, depending on the material that the light encounters. It is this wavelength dependence that gives objects the colors that we see. Sunlight contains light from every color in the rainbow, but when it reflects off of an object, it is only the reflected colors that we see. For example, paint on a red car absorbs most of the colors in sunlight but reflects back light in the red part of the spectrum, causing us to say, "That's a cool red car!" Colors that are not reflected are transmitted or absorbed. The absorbed light turns into heat energy, so black objects that absorb all colors are hotter in sunlight than are white objects, which reflect back all colors. Heat energy from the absorbed light is also radiated from the object as infrared light. While infrared light cannot be seen with our eyes, cameras designed to detect infrared light see the brightness of objects based on their heat rather than the amount of light they reflect back.

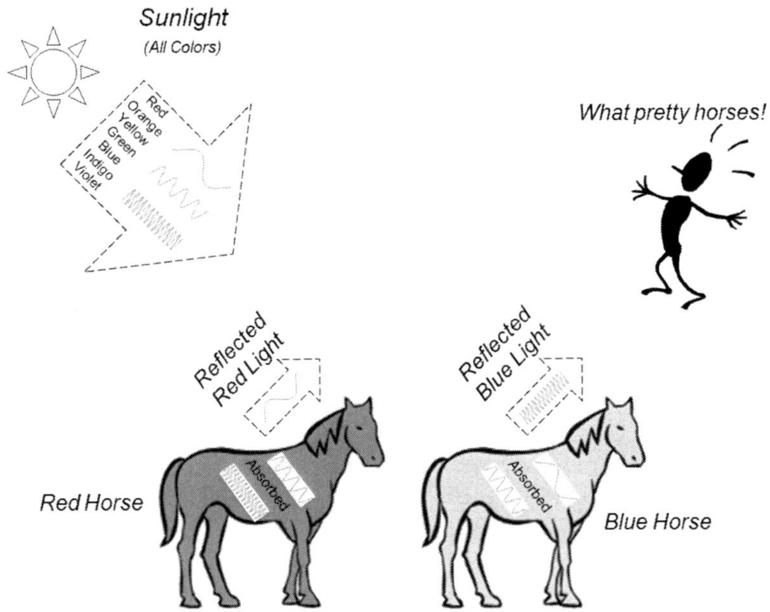

Colors of objects are determined by the colors of light that are absorbed or reflected.

Scientists take advantage of this wavelength dependence to classify materials based on the light spectrum that is reflected. Cameras designed to capture different colors reflected from objects are called spectral cameras. Multispectral cameras divide the spectrum into a few individual colors, generally ten or less, whereas hyperspectral cameras divide the spectrum into many individual colors, typically hundreds, to get better detail on the material's spectrum.

The colors that we see in objects depend on the wavelengths of the colors available in the light source. A red car won't look red if the light source doesn't contain red light because there will be no red light to reflect back to our eyes. We experience this effect when we look for our car at night in a parking lot that is illuminated with yellowish sodium vapor lights. Our beautiful car in the sunlight looks like a dark muddy color under these lights, so it can be hard to recognize. The color we expect to see is not possible under these lights because the colors necessary to reflect back our beautiful car's color in the sunlight are absent.

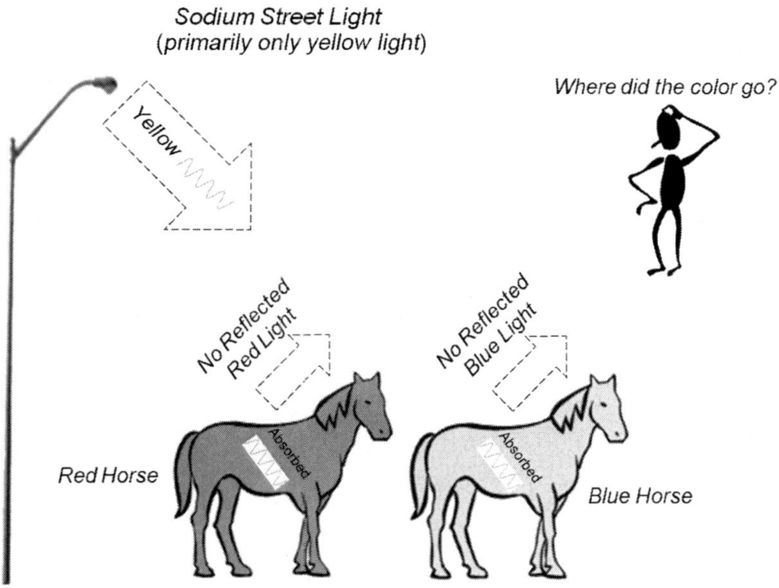

If the light source does not emit the necessary red and blue colors, we end up with colorless dark horses.

3.4 Light as the Camera Sees It

A camera designer needs to understand the properties of the light that will enter the camera; this means that the history of the light must be modeled in an imaging chain. For daylight imaging, light originates from the sun, which fortunately approximates a blackbody model, but then it interacts with the atmosphere when it arrives at Earth. The atmosphere is tricky to model because the properties of light passing through the atmosphere change with weather, humidity, pollution, and even solar flares. Light then reflects off the objects in a scene before entering the camera, and the specific characteristics of each object change the color and intensity of the light. To complicate the process even further, light that is scattered from the atmosphere and other elements in the scene also finds its way into the camera, generally producing a hazy effect in the pictures.

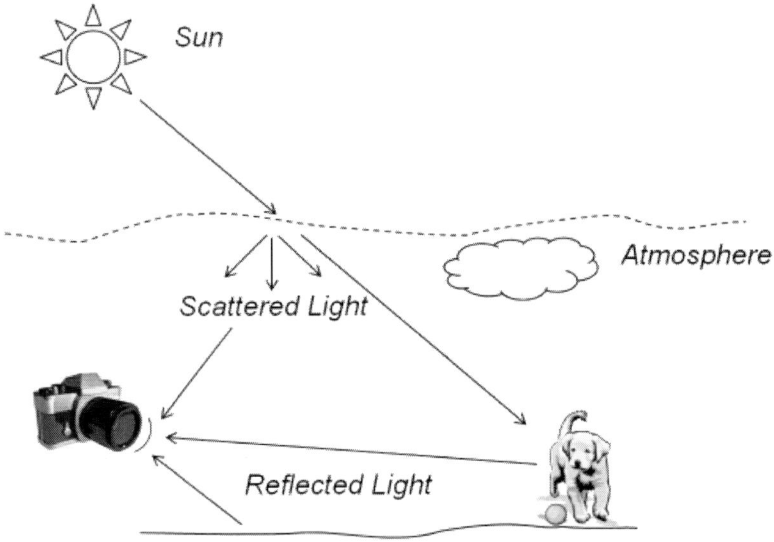

Light takes a complicated journey from the sun to the camera.

Getting a grip on light that enters the camera is important because this is the energy that creates the image; but even for a typical daylight photograph, this is a complicated process. Fortunately, years of research and development from clever scientists and engineers allow for reasonable predictions about the light energy that makes the picture.

Chapter 4

Camera Optics: Taking Control of the Light

4.1 Getting the Light into Shape

Optics is the study of the generation, propagation, and detection of light. In the camera-design world, camera optics refers to the elements of a camera that capture light to form an image. The role of the camera lens is to shape the light entering the camera into an image we wish to keep. Fortunately, the image will form naturally if we bend the incoming light in just the right way. Ideally, the image should be an exact copy of the scene in color, brightness, and geometry, but there are physical limitations that allow the image to be only a decent representation of the scene. So, how do we use optics to shape the light into an image?

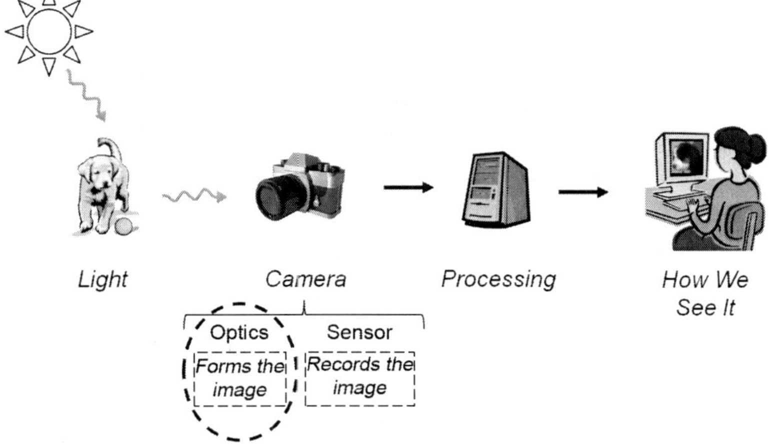

Optics forms the image in the camera from the light.

The easiest way is to let the light pass through a tiny pinhole, but as discussed previously, this severely limits the amount of light available to create the image. The hole needs to be wider to let more light in, but the image blurs if the light is not bent back into shape. The two most common ways to shape light are to use lenses, which use the law of refraction to redirect the light, or mirrors, which use the law of reflection to redirect the light.

4.2 Lenses

As described earlier, light can change direction when it passes through a transparent material such as glass. The light slows down as it passes through glass from air, so by changing the shape and thickness of the glass, we can control the direction of the light as it strikes different areas of the glass. If the piece of glass has a changing thickness, the light traveling through the thick part will fall behind the light traveling through the thin part, causing a change in direction. To help us understand how light passing through a wedge of glass changes direction, imagine a line of soldiers advancing through a large triangular patch of sand that is wider farther down the row of soldiers.

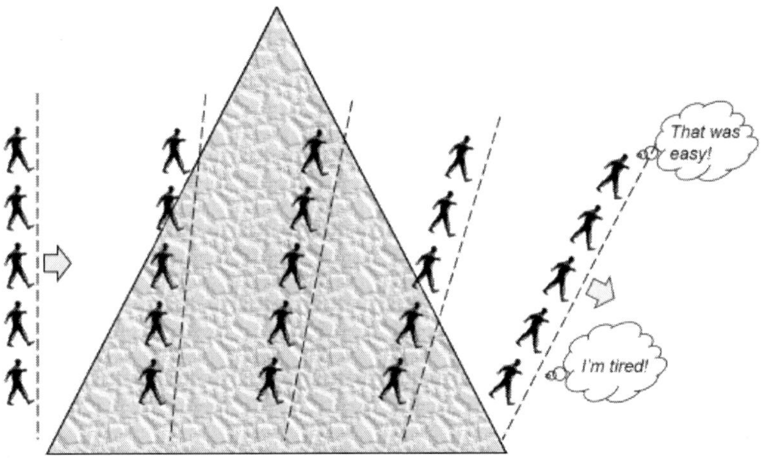

Light changes direction as it passes through a wedge of glass, just as a line of marching soldiers needs to change direction to maintain a straight line when they encounter a wedge of sandy ground.

We can assume that because it is very hard to march in loose sand, the sand will slow the soldiers down when they cross through it. The lucky soldiers that march through the narrow area of sand will not slow down very much, but the poor soldiers that encounter the wide area of sand will slow down quite a bit. To maintain a straight line, the faster soldiers will need to slowly turn their direction toward the slower soldiers until they have all marched through the sand. When the soldiers emerge from the sandy area, they will be marching in a new direction.

The shape and thickness of a piece of glass can be designed to make a lens that redirects light in a prescribed manner. The simplest lens is called a converging lens because it can be used to converge and bring together the light that enters the lens. A magnifying lens is a simple converging lens; this fact is discovered by many young children when they use a magnifying glass to converge sunlight onto an unfortunate ant.

It is easy to illustrate where light from a lens forms an image by using geometric optics to trace the direction of different rays of light as they pass through various parts of a lens. When all of the rays are parallel to the optical axis (the line that goes through the middle of the lens and is also the axis of rotational symmetry), the light is *collimated* and converges at a focal point. The image formed when all of the light is brought together is called a real image because light is actually there, and we can use this light to project an image onto a wall or onto an imaging sensor. The distance from the lens to the focal point is called the focal length, and this is usually the first information given to describe a lens.

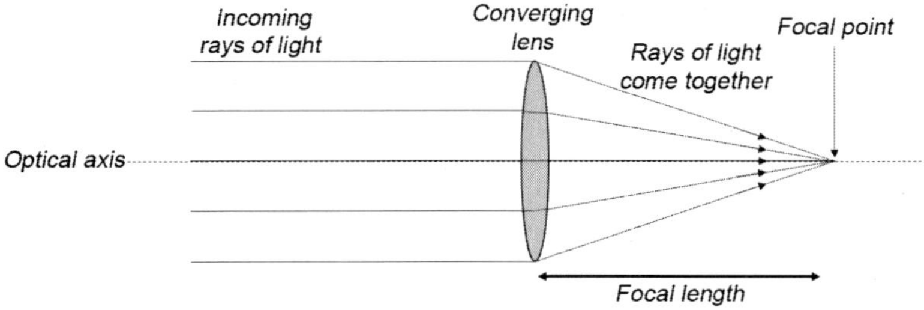

A ray-tracing diagram shows light coming together for a converging lens.

By contrast, a diverging lens is designed to diverge or bend light away from a common point. The focal point of a diverging lens is the point where the light seems to be coming from, even though it does not come from that point at all. The image formed by a diverging lens is called a virtual image because no light energy is actually coming together at the focal point, so an image cannot be projected or recorded.

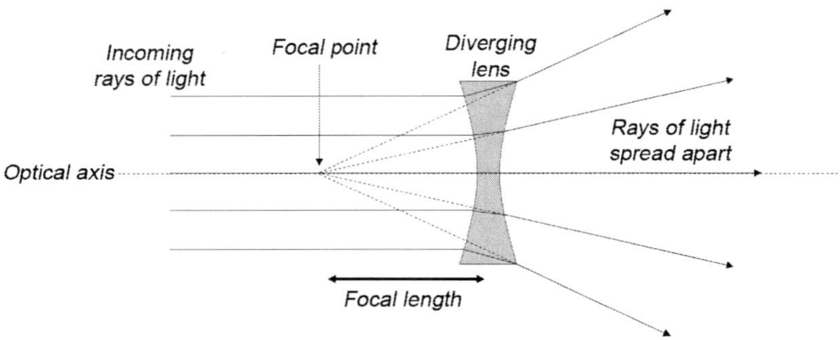

A ray-tracing diagram shows light spreading out for a diverging lens.

People with blurry vision typically suffer from a condition where the lenses in their eyes do not properly bring light together on the retina. This type of eyesight can be corrected by placing lenses in front of their eyes that redirect the light onto the retina. When the optometrist clicks through different lenses and tirelessly asks us, “Better or worse?” he or she is trying to find the shape of the lens that best compensates for the misgivings of the lenses in our eyes. People who are near-sighted (myopia) need a diverging lens to correct their vision, while people who are far-sighted (hyperopia) need reading glasses that are converging lenses. Keep this in mind if you are lost and you want to start a campfire because it is the far-sighted person who can use their eyeglasses to create the fire!

Near-sighted vision

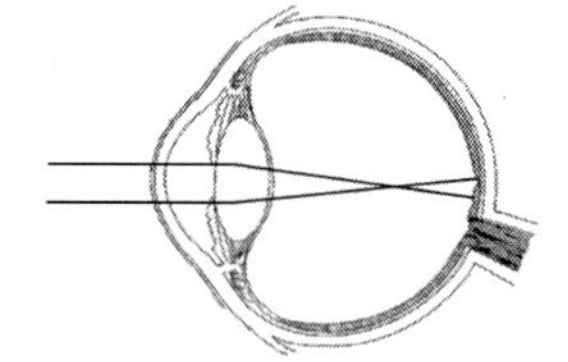

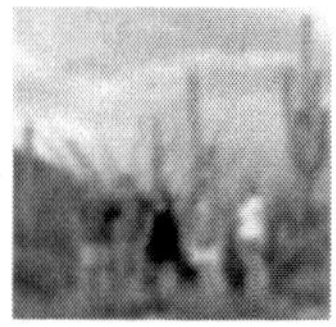

The image focuses in front of the retina.

Objects far away are blurred.

A diverging lens spreads the light out to focus on the retina.

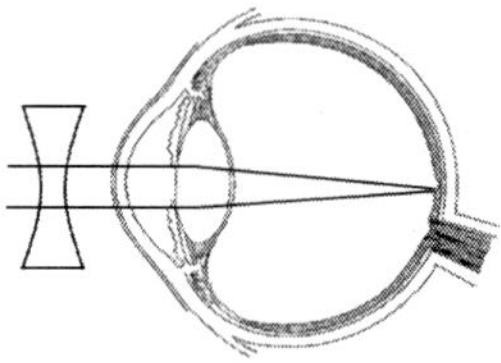

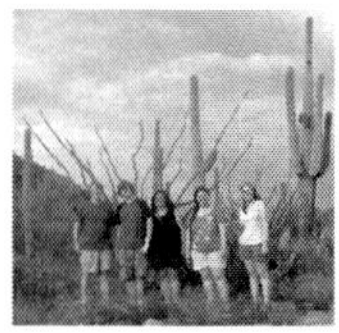

Near-sighted people need diverging lenses to correct their vision.

Far-sighted vision

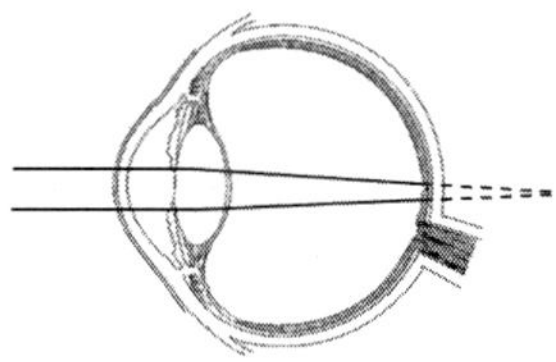

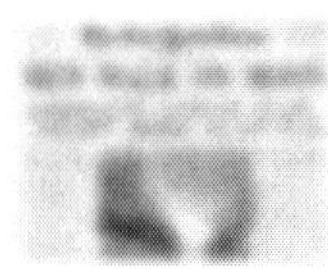

The image focuses behind the retina.

Objects near the eye are blurred.

A converging lens brings the light together to focus on the retina.

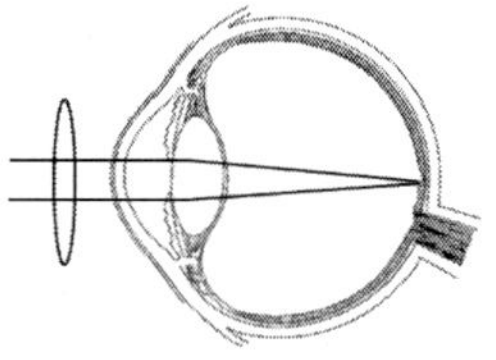

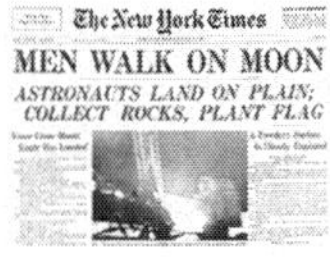

Far-sighted people need converging lenses to correct their vision.

One of the difficulties when using glass lenses to focus light is a property called dispersion. The amount of bending (or refraction) that occurs is dependent on the color of the light, so different colors do not bend equally. This causes the light from all of the colors to disperse and spread out, resulting in a blurred image. Another problem with glass lenses is that they quickly become heavy as they grow larger, due to the thickness required for the glass to bend the light. Although these issues are generally not a problem for hand-held cameras, they can be a big problem for astronomers who need to build big apertures to collect as much of the faint starlight as possible to form a high-quality image. Astronomers, therefore, turn to mirrors to redirect light using reflection rather than refraction.

4.3 Mirrors

When we admire our new haircut in the mirror, the mirror must be as flat as possible so that our reflection is not distorted. Any bending or curvature in the mirror distorts our image, which is great entertainment in a circus funhouse but not so much fun when we're admiring a new hairdo. A flat mirror does not bring light together to form a real image that can be recorded, so we need to curve the mirror in just the right way to form an image onto a sensor.

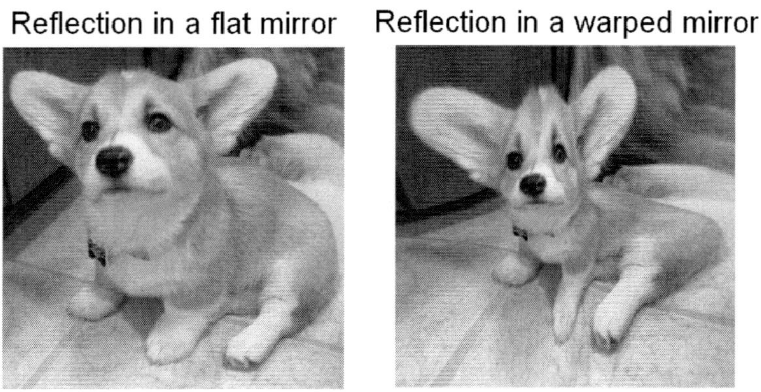

As we know from circus funhouse mirrors, the direction of reflected light changes with the curvature of the mirror surface.

The direction of the curvature on a mirror is described as either concave or convex. A concave reflecting surface, such as the inside of a metal mixing bowl, curves toward a scene and can bring light together to form a real image, similar to a converging lens from which we can record an image. A convex reflecting surface, such as a shiny ball bearing, curves away from a scene and reflects the light away to create a virtual image, similar to a diverging lens. This type of mirror by itself would not help us to form an image when making a photograph.

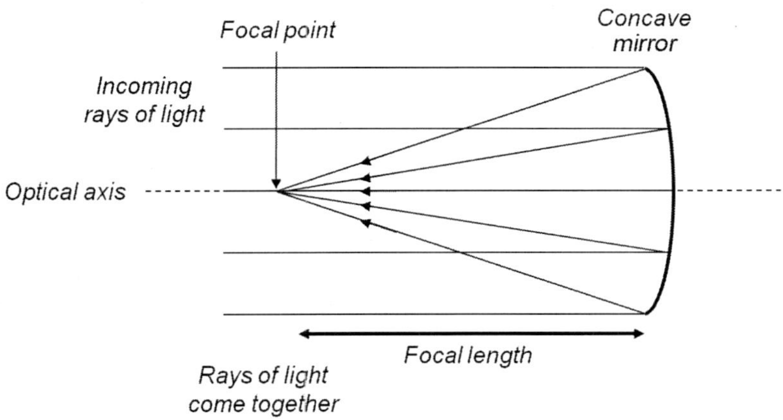

A ray-tracing diagram shows light coming together for a concave mirror.

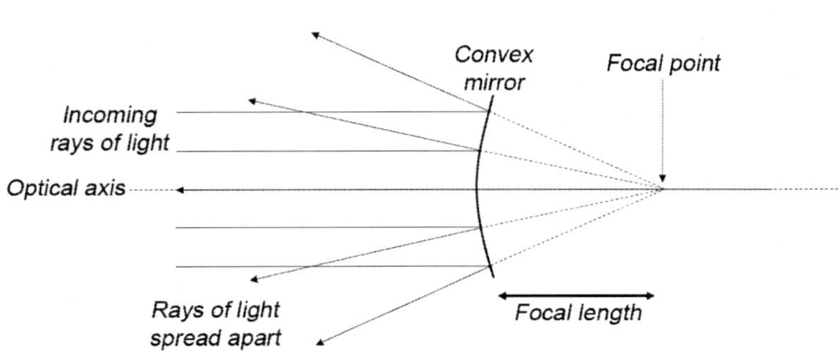

A ray-tracing diagram shows light spreading out for a convex mirror.

Telescopes used by astronomers are usually designed using a second mirror to reduce the length of the telescope. A large primary mirror collects light and brings it together to form an image, but a secondary mirror is used to reflect the light back toward the primary mirror before an image is formed. (This design is called a Cassegrain reflector, which first appeared in a publication on April 25, 1672. It's interesting to note that the identity of the true inventor is still a mystery, but it is generally attributed to the Catholic priest Laurent Cassegrain.) Now that light is heading back toward the primary mirror, a hole needs to be cut in the primary mirror so that the light can pass through to reach a sensor. Some incoming light will be blocked as it encounters the secondary mirror. Although blocking the light this way takes away some of the light energy and adds a little blurring to the image, in the big picture (pun intended!), these degradations are minimal to the overall image quality.

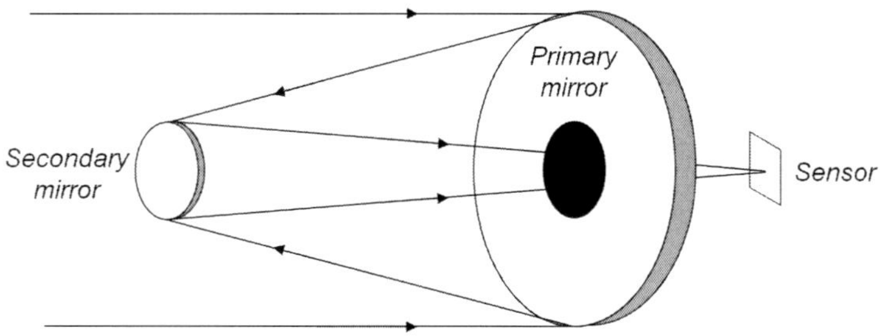

Telescopes with a design such as the Cassegrain typically use a secondary mirror to fold back light and shorten the length of the telescope.

4.4 *f*-Number Basics

The *f*-number (or *f*/#) is the ratio of the focal length to the diameter of the aperture opening where light enters the camera. For a camera with optics having an *f*-number of 8, the *f*-number is written as *f*/8. As simple as it is, the *f*/# is a very important parameter to optical designers. The blurriness and brightness of an image are dependent on the *f*/# of the camera. Another way to think of the *f*/# is the angle at which the light at the edge of the aperture comes into the sensor when it is focused to form an image.

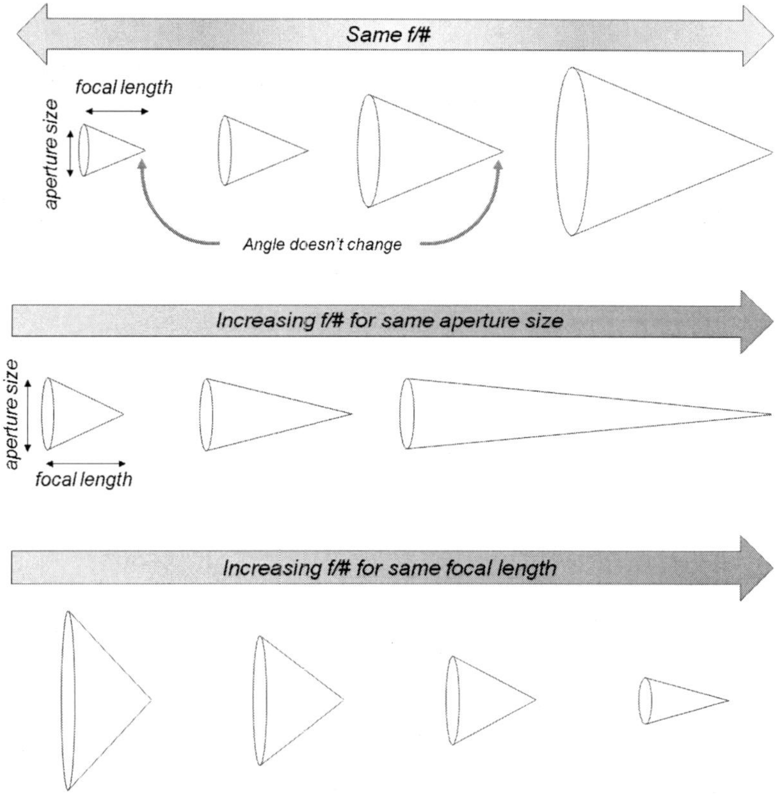

The *f/#* changes with aperture size and focal length of the optics.

One important aspect of the *f*/# is a property called the depth of field. When we focus our camera on an object in a scene, the depth of field is the distance in front of and beyond that object that will not show out-of-focus blurring. A high *f*/# will have a high depth of field, so objects in the foreground and background will all be in focus. A low *f*/# will blur objects in the foreground and background and is used by many photographers to give photographs the appearance of depth.

In the past, photographers usually knew the *f*/# as the *f*-stop because cameras used to have a fixed focal length lens on them, so changing the *f*/# was accomplished by *stopping down* or decreasing the entrance aperture. (Most cameras today have a

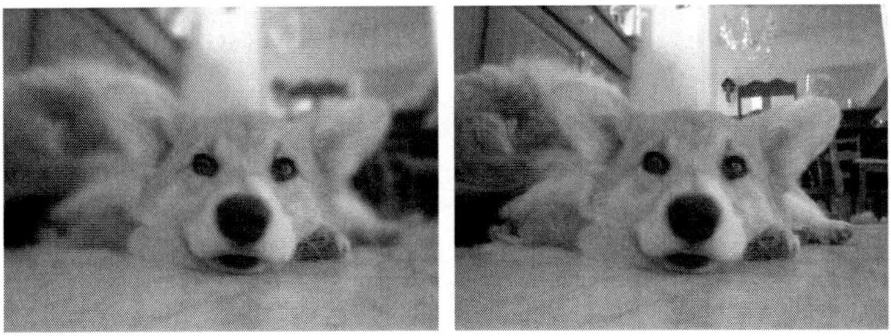

f/4 *f*/16

Small depth of field Large depth of field

A smaller *f*/# will blur more of the foreground and background in the scene.

zoom lens, which changes the focal length to zoom in on the scene.) The amount of light entering a camera is related to the area of the aperture, which is related to the square of the aperture diameter. So, to cut the amount of light in half, the aperture diameter must be reduced by a factor of the square root of 2. This is why *f*-stops on cameras seem to be in unusual increments (*f*/1, *f*/1.4, *f*/2, *f*/2.8, *f*/4, *f*/5.6, *f*/8, *f*/11, *f*/16, *f*/22, *f*/32, etc.), but they are actually a sequence of the powers of the square root of 2 such that each step down will double the amount of light entering the camera. Low *f*/#'s place more light onto the sensor, allowing faster exposure time, hence low-*f*/# systems are referred to as fast optics, while higher *f*/#'s are referred to as slow optics.

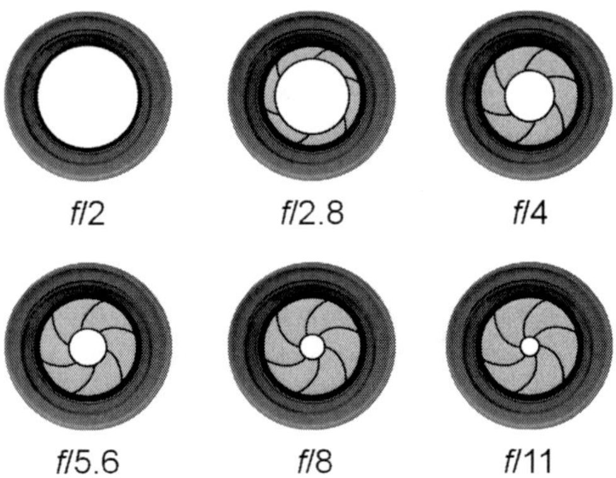

Stopping down the aperture of a camera with a fixed focal length changes the *f*/#. (In this example, the wide-open aperture is *f*/2.)

4.5 Forming the Perfect Image Is a Tall Order

Ideally we would like an image formed by the optics to be an exact copy of the scene in every imaginable way. To replicate the scene, we would need to record the three-dimensional aspects of the scene with the proper color and brightness. Then, there is the changing nature of the scene over time that would need to be recorded as well. We can imagine a future where we could walk through an image of a scene and think that we were walking through the actual scene, but for now we will discuss the limitations of the digital cameras that we use today.

Optical engineers work hard to design lenses and mirrors that bring all of the light together in just the right way to achieve a highly detailed copy of a scene. Unfortunately, this is harder than it seems. This is why lenses sufficient to form reasonably good images were not produced until the 16th century. The way light comes together to form an image is very sensitive to the shape of the lens or mirror, and very small changes can lead to noticeable differences in the quality of the image. These deviations from the ideal optical system are called aberrations, and optical designers take great pains to minimize them. They do so by adding more optical elements and creating optics with more complex shapes.

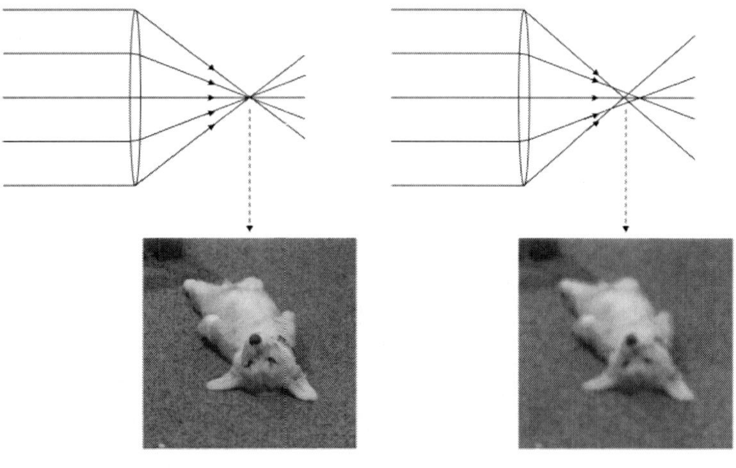

We see blurring in our pictures if light isn't brought together in just the right way.

The main effect of aberrations is the blurring of an image, as we have all seen when the image isn't focused properly. To get an idea of this sensitivity, if the lens in your digital camera misses the location where the light should come together by a distance of about ten hair widths or more, you will likely see a noticeable blurring in the image.

How can we measure how well our camera optics is pulling the light together? A simple way is to see how well it can record an image of a simple point of light. After all, if the camera can faithfully reproduce a tiny point of light in a scene, it can reproduce any detail in the scene on the scale of that tiny point. Using the ray-tracing approach that we've used so far gives the impression that all we need to do is properly bring all of the light coming through the optics together at the focal point, and voilà, we will have a high-detail image! Unfortunately, the ray-tracing approach does not take into account that light is an electromagnetic wave. We must consider the wave properties of light through the optics to understand how well the light is pulled together. Physical optics is the study of the wave nature of light and shows us how a lens or mirror changes the wavefronts of the light as they move through a camera.

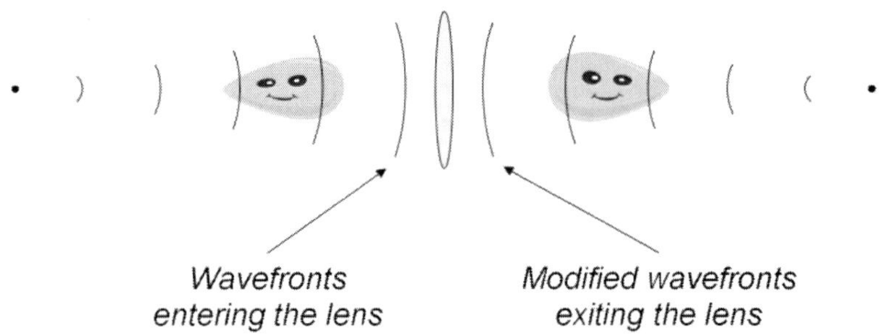

Physical optics shows us how a lens modifies the wavefronts of light to create an image.

Even with the highest-quality optics, the wave nature of light will prevent us from making a point-for-point copy of the scene, so the goal of imaging a perfect point is already lost. Why is the dream dashed so soon? Because as light travels, it spreads out, and the camera aperture can never be large enough to capture all of the light necessary to image a perfect point. We need the optics to form an image, but we can't make the size of the optics infinitely large (remember, this isn't Hollywood). When the light hits the edge of the camera aperture, it diffracts or spreads out from that edge, causing a point of light to go from skinny to fat. This means that a point of light in the scene spreads out into a blob by the time it is imaged in the camera. As we will see later, there are other factors in the camera that also cause a point of light to spread out by the time it is recorded in the image. This spreading of the point is described by the point spread function (PSF) of the camera.

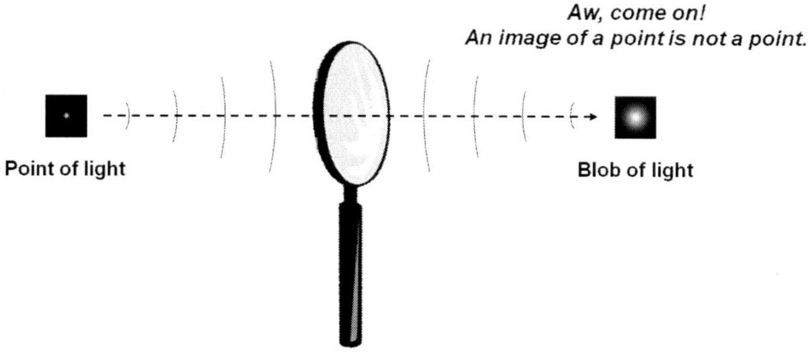

A point of light is not actually imaged as a point because of the wave nature of light and the limited size of light-collecting optics.

So, how do we calculate what the light will look like after it enters a camera aperture? Given a camera design, how do we know what the PSF will look like? What differences in the PSF are important? These are important questions because we want to make the PSF as small as possible so that a point of light in the scene will look as much as possible like a point of light in the image. But when is the PSF good enough? Mathematics for calculating the PSF and its effect on an image can be fairly difficult, but fortunately our work can be simplified by looking at image formation in a different light (pun intended, again!).

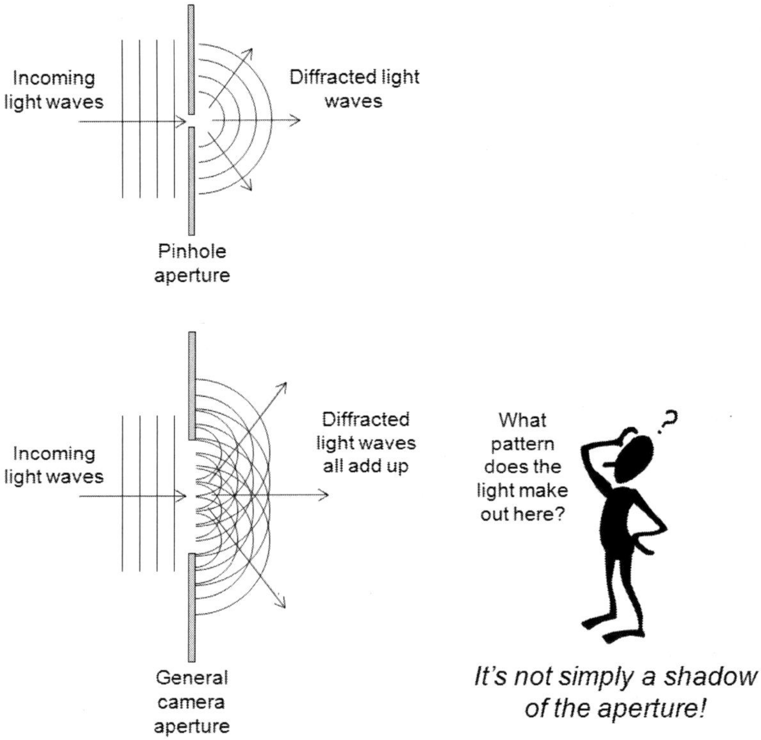

Light diffracts after passing through a camera aperture, further complicating the effort to understand the quality of the image.

Chapter 5
A Simpler Math

5.1 Don't be Frightened by LSI!

This is the chapter most readers will probably want to skip, but for those who are curious about the geek speak used by camera designers, press on! Understanding how light from a scene transforms into an image through a camera can be simplified if we think of the camera as a linear shift-invariant (LSI) system. That's a bit of a mouthful, but the concept is pretty straightforward, so let's look at the first part to understand a linear system. When an operator, say a dog, encounters an input, say a biscuit, we get a response—in this example let's say it's a "woof." If the operator encounters another input, say a cupcake, we might get a different response—this time let's say it's a "bark." But now we want to combine the inputs, so in this example we give the dog a biscuit and a cupcake. If the dog's response is simply the combination of the two responses given for the individual biscuit or cupcake, a "woof" and a "bark," we have a linear system. In other words, the response of a linear system to the sum of inputs is simply the sum of the individual responses. Easy, right?

Now let's look at the concept of a shift-invariant system. As you may have guessed, the form of the response won't change (it is invariant) when we shift the location of the input. If we move the dog to another location but still receive the same response, such as a "woof" when he gets a biscuit, we have a shift-invariant dog. Put this behavior together with the fact that the dog gives us a linear response and voilà, we have a linear shift-invariant dog!

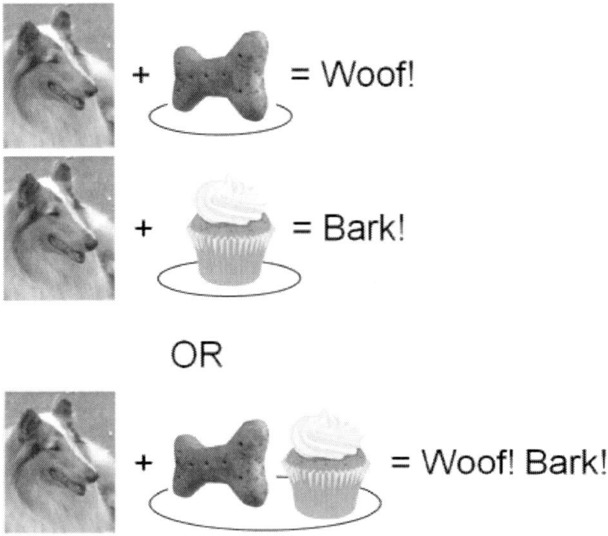

If two inputs together simply give the combined response of each individual input, we have a linear system.

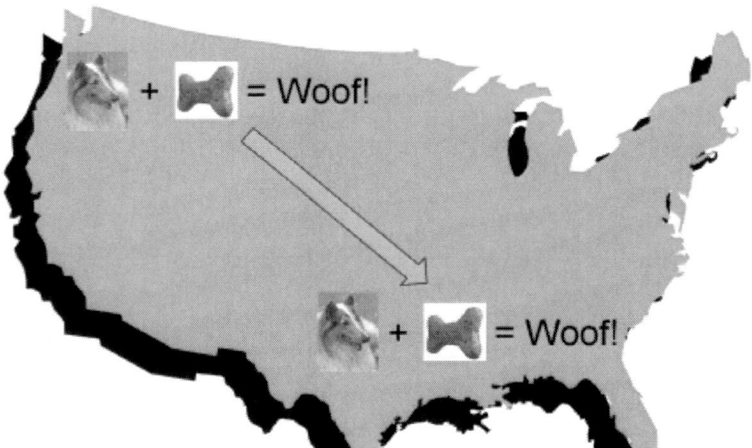

If the response is the same regardless of location, we have a shift-invariant system.

But what does this have to do with forming an image? Well, if we know how much a point of light is blurred by a camera (remember this is called the PSF), and if *every* point in the scene becomes blurred into the same shape (although the brightness of the blurred point scales with the brightness of the point in the scene), the camera is a linear shift-invariant system. This means that we can blur the entire *image* by simply knowing how a *point* becomes blurred (described by the PSF) and applying that PSF to all of the points in the scene. This saves us the tremendous amount of time it would take to calculate the blurring of every point in the scene by calculating how the light propagated at each point. This, of course, frees up valuable time to play with a puppy.

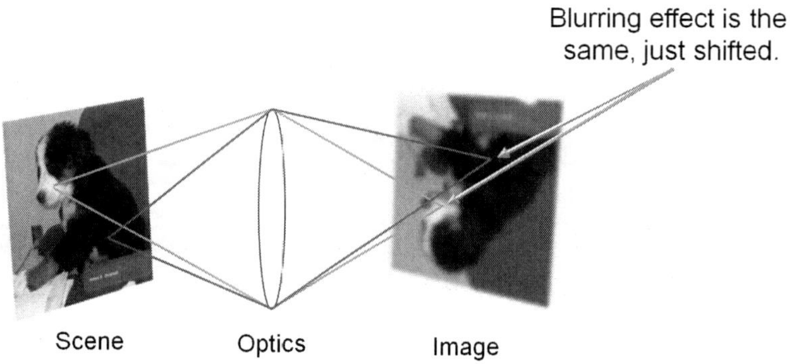

If the blurring effect from the optics is the same everywhere, we have a linear shift-invariant (LSI) system.

The process used for applying the PSF to the scene is called a convolution and is usually represented with an asterisk symbol (*). We can think of smudging the image with the PSF using a shift–multiply–add procedure that smudges the image one pixel at a time, from left to right and top to bottom, resulting in a blurred image. If you smudge a drawing with a big thumb instead of a small pinky finger, you will see more smudging. Similarly, a bigger PSF will blur an image more than a smaller PSF will, so camera designers work very hard to keep the PSF of the camera as small as possible.

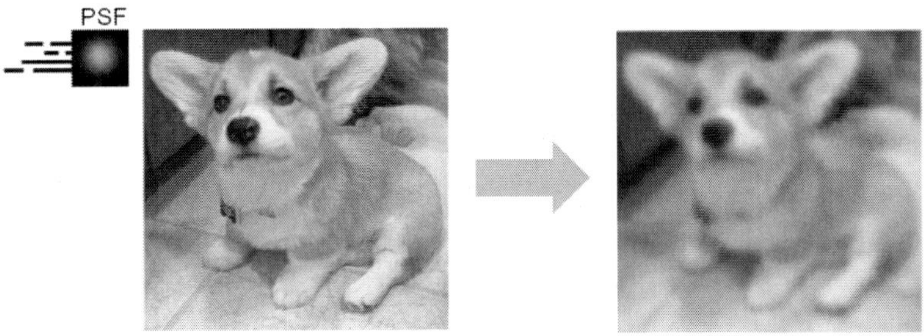

The convolution process blurs a scene by using the PSF to mix together neighboring pixels.

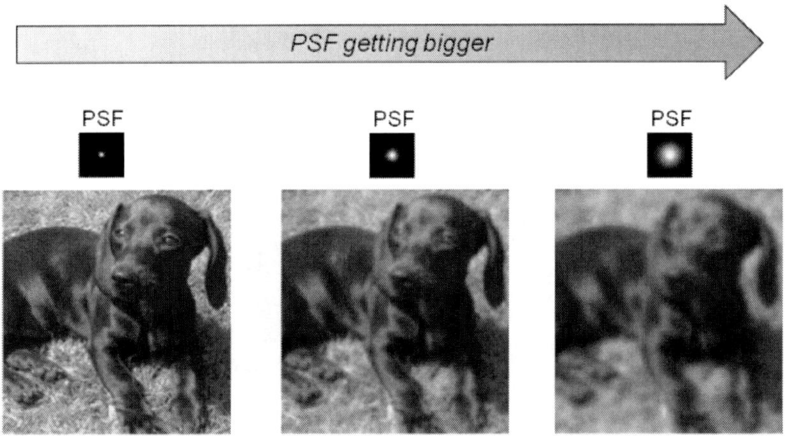

A camera with a larger PSF produces a blurrier image.

There are, unfortunately, two burning concerns that we need to address if we want to understand and characterize the blurring produced by a camera. The first concern is determining to a high level of detail what the PSF looks like. What is the shape and how big is it? Small changes in the PSF can have a visible effect on image quality, so we want to get it right. The second concern is the time it takes to conduct the convolution process if we wish to apply it to an image to simulate the blurring that a camera will have on an image, especially if the PSF and image are large. Fortunately we can address both concerns with a favorite topic of camera designer, Fourier transforms.

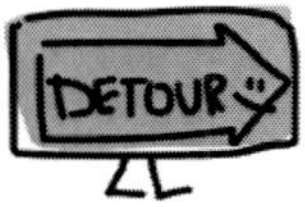

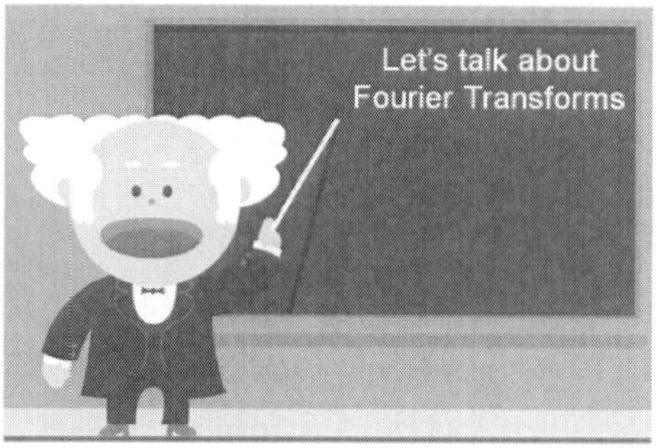

We can understand the blurring of an image better if we talk about Fourier transforms.

5.2 A Fourier What?

People who work with camera designers and image scientists hear them talk about Fourier (pronounced foor-ee-ay), but they never explain their obsession with this word (or if they do, you usually wish they hadn't). The term comes from the French mathematician Jean Baptiste Joseph Fourier (1768–1830), who saw things a little differently from the way other mathematicians of his time saw things. Fourier introduced the idea that mathematical functions could be described by adding together different waves. In the world of imaging, Fourier would see everything as the combination of many different waves.

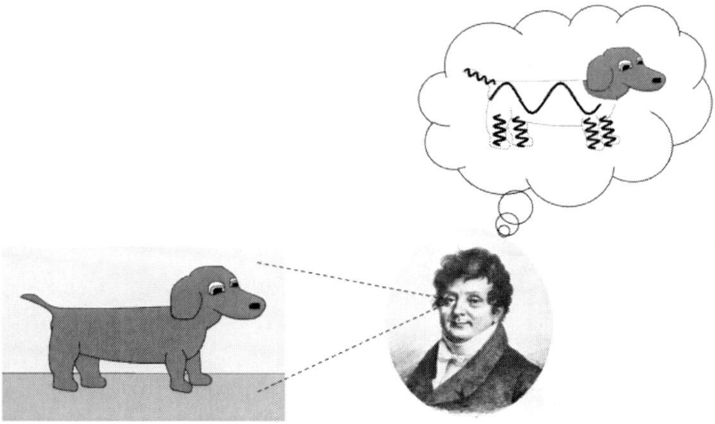

Fourier saw the world in terms of waves.
(Dog illustration courtesy of Katie Fiete.)

You are probably saying to yourself, "Life is hard enough already, why would we want to think of an image that way?" It turns out that relating a camera design to the quality of the image is much easier if we think of an image in terms of waves. What waves, you ask? Why spatial waves, of course! Unlike electromagnetic waves that describe the propagating light, a spatial wave can be thought of as periodic bright and dark patterns across an image. The wavelength of the spatial pattern is the distance between neighboring bright bands. It's actually more convenient if we think of the waves in terms of their spatial frequency instead of their spatial wavelength. For example, if a wavelength is measured in meters, the spatial frequency will be measured in the number of cycles per meter.

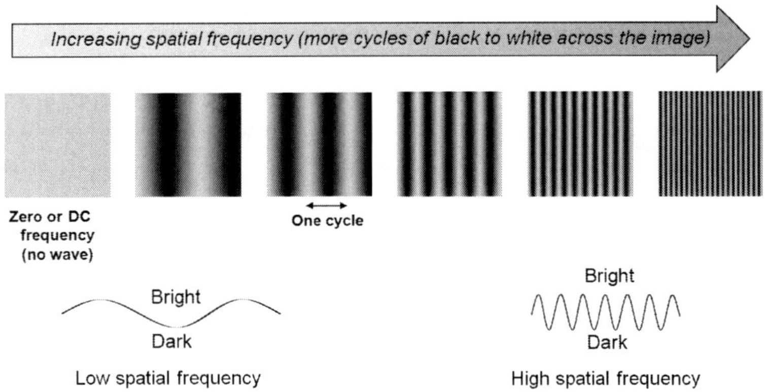

Spatial waves in an image look like stripes that fade between black and white. Waves with higher spatial frequencies cycle more times between black and white across an image than do waves with lower spatial frequencies.

So, how do we think of an image in terms of waves and spatial frequencies? Fortunately, the image can go through a mathematical transformation, called a Fourier transform. This transformation turns the image into another image, which gives us the recipe for making that image out of waves. We probably all have a recipe for making a cake out of sugar, flour, and other ingredients, but what is the recipe for making an image of a cake out of waves? Ah ha! The Fourier transform tells us!

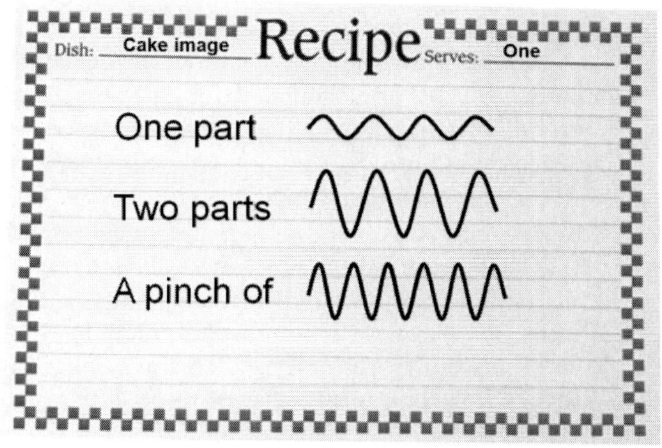

The Fourier transform gives us the recipe for making an image out of waves.

We need to know how much of each type of wave to add together to achieve the desired image. Adding the wrong amount of an ingredient could result in something quite different and probably not to our liking. The spatial waves are characterized by their spatial frequency (or wavelength), amplitude (strength), orientation, and phase (how much they shift). The Fourier transform is a process that transforms an image into a recipe that tells us how much of each type of wave to add and how much to shift them.

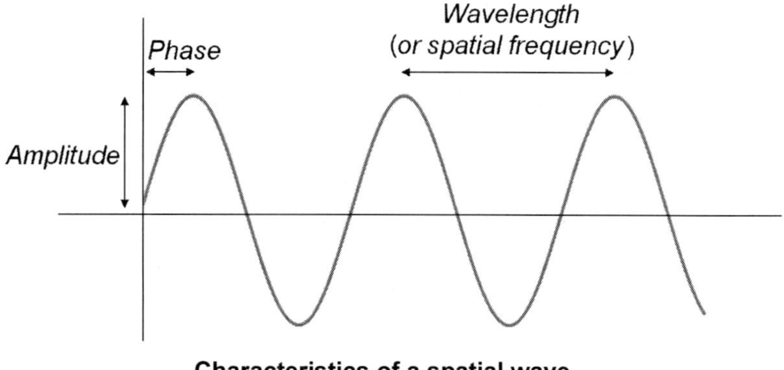

Characteristics of a spatial wave.

The frequency of each wave is described as a *spatial* frequency, because it relates to the spatial makeup of an image. Low spatial frequencies are the broad parts of an image without much detail, such as the background. High spatial frequencies are the parts of the image with high detail, such as the edges.

The Fourier transform of an image, called the *image spectrum*, gives us another image that is a pictorial representation of the recipe for making the image by adding waves together. Unfortunately, we cannot display the whole recipe in one image because there are two parts, called the *modulus* and the *phase*, so typically only the modulus image is shown. The modulus shows how strong each wave needs to be in the recipe (like adding a lot of paprika, or just a little), whereas the phase shows how the wave needs to be shifted around before it is added.

We can think of the image of the modulus as a map of all of the spatial frequencies in an image; the brightness of each location tells us the amplitude of a wave at that spatial frequency that needs to be added to the mix. Typically, the image is displayed with the lowest (zero) spatial frequency in the middle and the highest possible spatial frequency at the edge. The image of the modulus may not look very interesting, but it's worth noting that the lower spatial frequencies are typically brighter than the higher spatial frequencies because the predominant

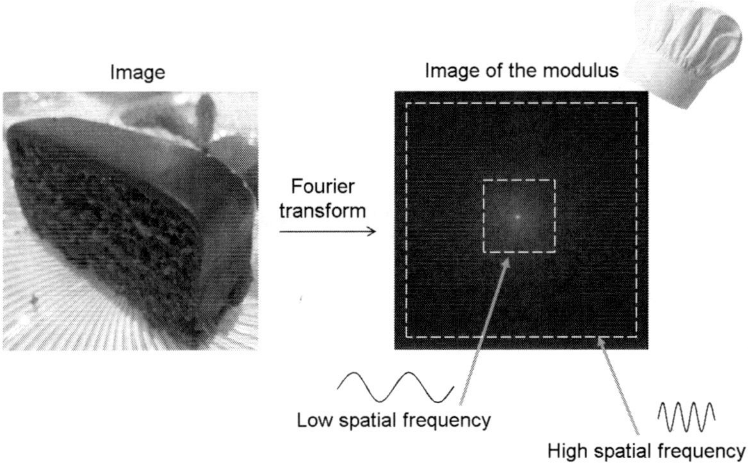

The recipe for making an image of a cake by adding together waves is given by the Fourier transform.

features in most images are big, broad objects, such as the background, so we need a healthy dose of those waves.

5.3 Modulation Transfer Function, or How I Learned to Blur an Image

Now that we are experts in making Fourier recipes for images, how does that help us? We know that the details in an image are in higher spatial frequencies, so it's convenient to see how cameras preserve or destroy them. The function that describes how spatial frequencies are modified by a camera is called the transfer function. Better yet, the transfer function is the Fourier transform of the PSF. (I can tell you're impressed!) And even better, camera designers love to work with transfer functions because the blurring of an image can be performed as a simple multiplication with a transfer function rather than a convolution with a PSF. Most of us would rather deal with multiplying than convolving.

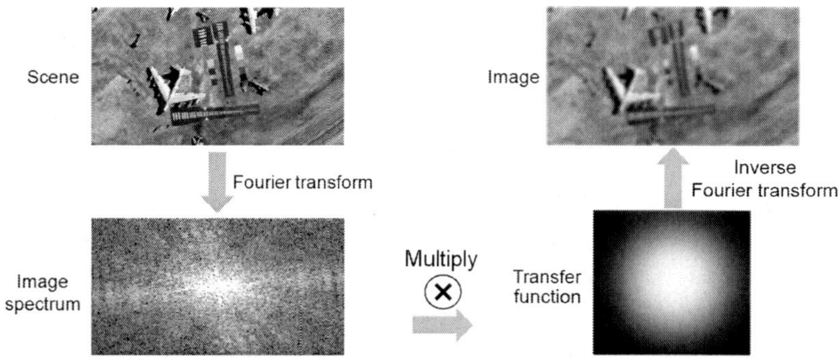

Blurring an image is simple multiplication when we use Fourier transforms and transfer functions.

It's important to note that a wide PSF produces a skinny transfer function, and a skinny PSF produces a wide transfer function. Remember that a wider PSF blurs an image more than a skinny PSF does, so we want the camera's PSF to be as skinny as possible. This means that skinny transfer functions blur an image more, and wide transfer functions blur an image less; so a wide transfer function is needed to pass the higher spatial frequencies necessary to prevent blurring. Put another way, the small skinny details in an image need a fat transfer function to prevent them from being blurred away.

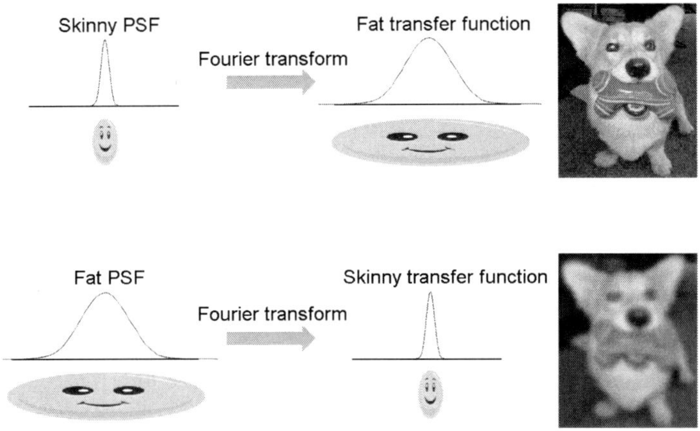

An image will be blurred if a camera has either a fat PSF or a skinny transfer function.

Just as the Fourier transform of an image has two parts—a modulus and a phase—the transfer function has a modulus and a phase as well. And just as we typically show only the modulus of an image spectrum, we also typically show only the modulus of a transfer function, called the modulation transfer function (MTF). The MTF is very useful because it shows how the contrast of each spatial wave is changed at each spatial frequency, relative to a wave with a spatial frequency of zero (remember, this spatial frequency represents a flat background). An MTF has a value of 1 at a zero spatial frequency, and it usually drops down for higher spatial frequencies, reducing their contrast and causing the blurring effect we see in images. So when a camera designer starts yelling, "I need a better MTF!," what he is essentially saying is that he wants less blurring.

An ideal system has a value of 1 for the MTF everywhere because this means that the system is not changing the contrast of any spatial frequency. Rather than showing the MTF in the form of an image, it is typically shown as a plotted curve. The MTF is usually symmetric, meaning that the left side is merely a reflection of the right side, so only one side of the curve is usually shown. Remember, Fourier said that a scene is made up from different spatial waves, and the MTF shows how the system changes the contrast of each spatial wave. The MTF curve for a typical camera starts high and then tapers down toward zero. The faster the curve drops toward zero, the more the image blurs.

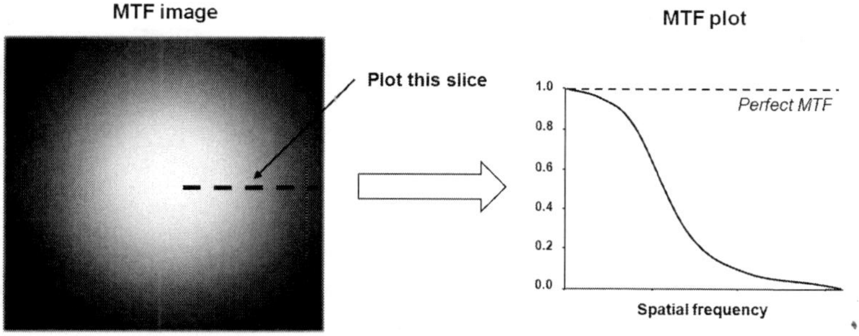

An MTF plot is a quick and easy way to illustrate how much the camera is blurring an image.

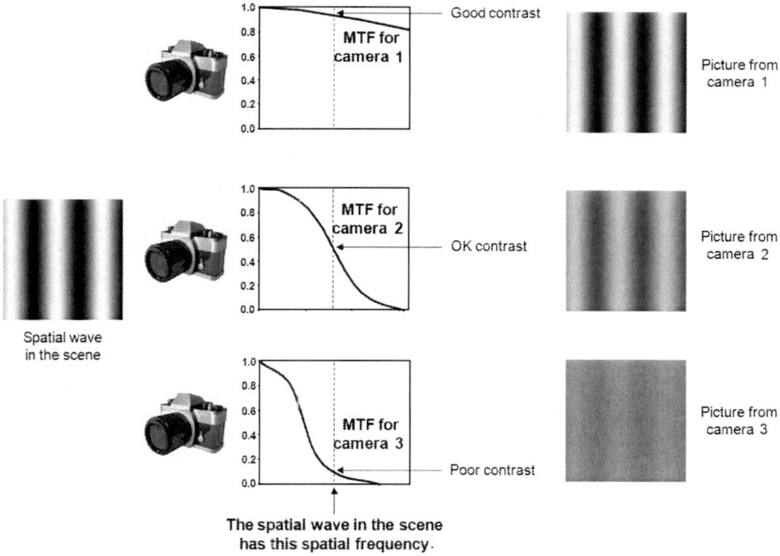

A lower MTF curve means that a spatial wave has less contrast, which looks like blurring in the image.

A chirp image is a neat way to look at the blurring that an MTF causes in an image. A chirp image is an image with a wave pattern that increases in spatial frequency as we move across the image, so we can see the effect that the MTF has on all of the spatial frequencies by looking at one image. An image with increasing spatial frequency is called a chirp image because a sound that increases in frequency will sound like a bird chirp that starts low and ends high. You can think of this as seeing a sound! If we applied an MTF curve to a bird chirp, the sound would fade away as the pitch got higher, with the sound fading away quicker for an MTF that curves down quicker.

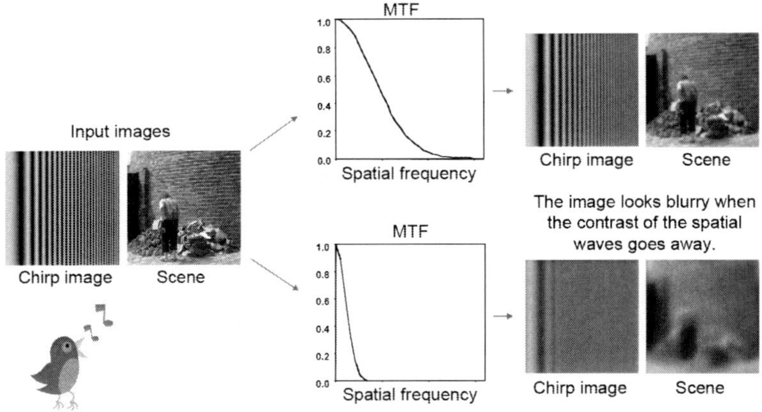

A chirp image is a neat way to see the effect of a transfer function on spatial waves with different frequencies.

Each part of the imaging chain can cause additional blurring in an image. For example, the optics, sensor, and motion of a camera each contribute to the blurring of the final image. So, how do we put together the blurring from each element of the imaging chain to determine the overall system blurring that occurs in the final image? This is the best part, at least if you're a camera designer—the blurring from the entire camera system can be determined simply by multiplying the transfer function from each element of the imaging chain.

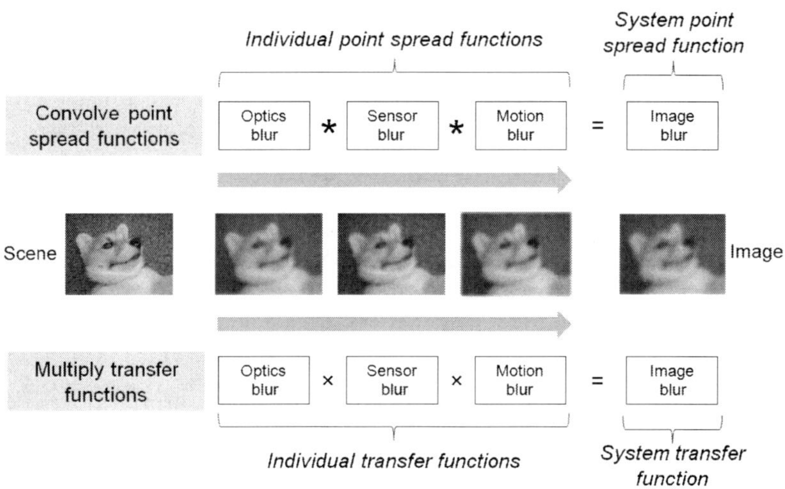

Image blur from a camera can be determined simply by multiplying the individual transfer functions.

Now that you know how exciting transfer functions can be, there is one more neat property. We saw earlier that the diffraction of light from a camera aperture causes the light to spread out in such a way that a point of light will be imaged as a blob described by the PSF. But how does the aperture influence the shape of the PSF? By convolving the aperture shape with itself, we achieve the transfer function, which gives us the PSF through the inverse Fourier transform. Pretty slick! By shining light through apertures of different shapes, the image of a point of light will take on different patterns. The most common aperture shape for most camera designs is a simple clear circular aperture that gives rise to a PSF, called the Airy pattern, named after the English mathematician and astronomer George Biddell Airy (1801–1892).

Let's look at the effect on the PSF when we block part of a circular aperture. In particular, let's look at the aperture of a Cassegrain telescope that requires not only a smaller secondary mirror in front of the bigger primary mirror, but also struts to hold that smaller mirror in place. Blocking part of the aperture

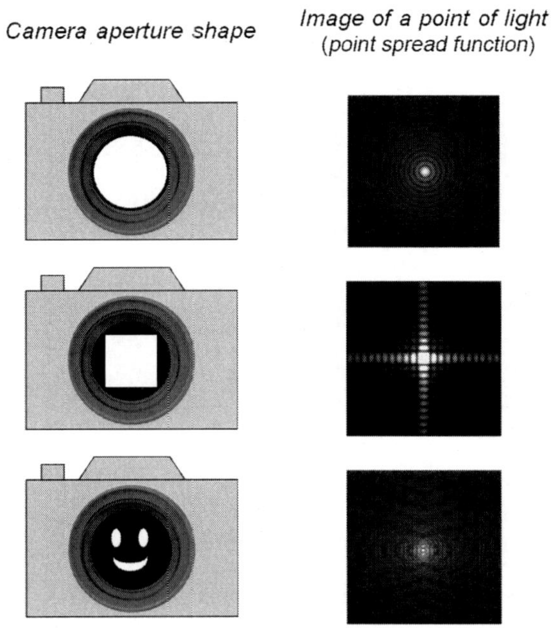

A camera's aperture shape greatly influences how the image of a point will look. (Note that the PSFs shown here are for a single wavelength of light.)

doesn't sound ideal, right? It does change the shape of the PSF, but it does not affect the highest amount of detail that the telescope can possibly see. The secondary mirror blocks the light entering the telescope and adds a circular obstruction to the aperture, causing an additional pattern of rings in the PSF. The secondary mirror cannot just float in front of the camera, so long arms of structures called struts are used to hold it in place, but these struts further block the light. If four struts are used, the PSF takes on a cross pattern. Finally, if we add up the PSF for each wavelength of light over the visible region, the ring pattern smooths out in the final camera PSF. This example is particularly interesting because now we can understand where the cross patterns come from that we see in the beautiful star images taken from the Hubble Space Telescope. The cross patterns show up in many illustrations of stars today, but they are not really there; it's simply how a telescope made with four struts sees a star. And now we know this because we're getting smart about the imaging chain!

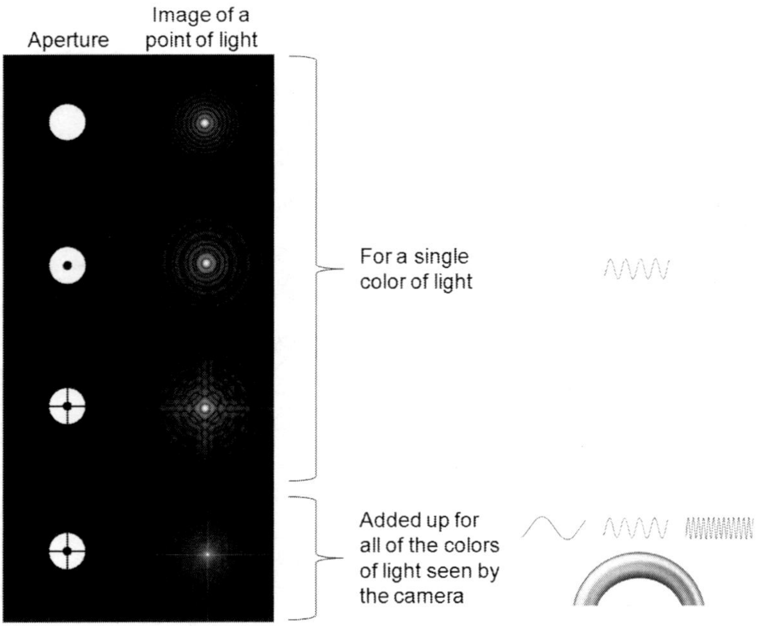

Adding obstructions to a camera aperture changes the image of a point of light.

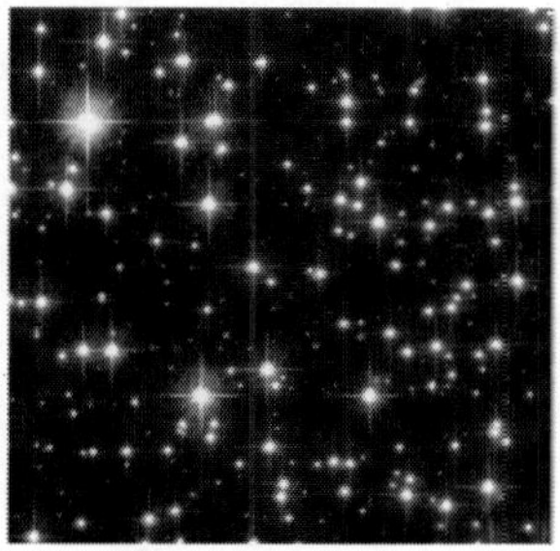

Star image by the
Hubble Space Telescope

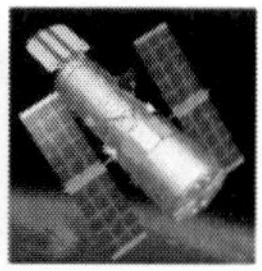

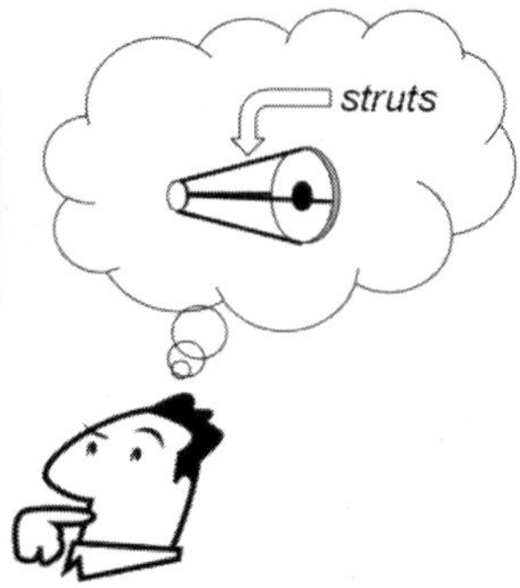

Struts holding the Hubble Space Telescope's secondary mirror in place create the cross pattern that we see on the stars in the photos. (Hubble image courtesy NASA, ESA.)

Chapter 6
Sensing the Light

6.1 Turning Light into a Digital Image

After the optics has formed an image, we need a way to record the image so that we can keep those bad-hair-day memories alive. An image is recorded using the aptly named imaging sensor because it senses the light and converts it into a form that can be recorded. Film was the traditional sensor for cameras, but it is quickly becoming one of those things that grandparents talk about: "When I was your age we used something called film!" For those young readers who have never used film, it is worth noting that film contained light-sensitive silver-halide crystals that blackened when exposed to light. Although film served well as a sensor, it was inconvenient because it had to be kept in the dark to prevent unwanted exposure, and the images couldn't be

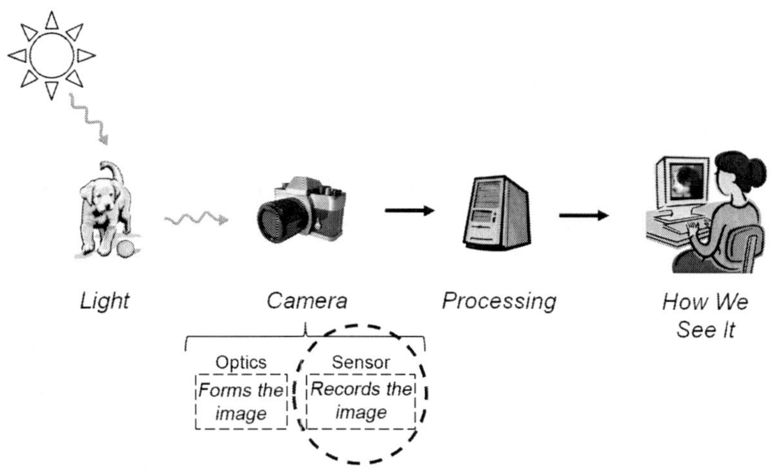

A sensor records the image formed by the optics.

seen until the film was treated with chemicals to produce the developed film negative. It was a negative image because a brighter light turned silver-halide crystals darker; thus the dark objects were light and the light objects were dark. The unexposed film was transparent, so the regions that grew darker from exposure became less transparent. A positive image was formed by shining light through the developed film onto paper treated with silver-halide crystals, so the bright areas of the negative (that corresponded with the dark regions in the scene) turned the paper dark. Most cameras today use digital sensors, which provide real-time electronic images that can be shared and stored on a computer quickly and easily, and without all that mess.

6.2 Making Numbers

The most common sensors used in digital cameras are either charge-coupled-device (CCD) sensors or complementary-metal-oxide-semiconductor (CMOS) sensors. The difference between CCD and CMOS sensors is in the electronics, but they both use a rectangular array of detectors to convert light to an electrical signal. Assuming that the digital image has not been resized, each detector in the sensor will represent a pixel in the digital image, so it's important to understand what is happening in each detector that gives us that piece of the picture.

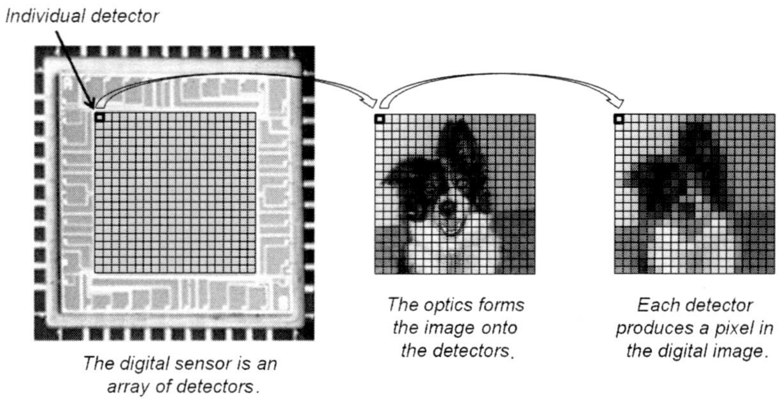

Each detector gives us a piece of the picture called a pixel.

As promised, we are going to look at light as a photon rather than a wave of electromagnetic energy. Why? Because the energy associated with each photon packs a punch that can kick an electron free in the detector through a phenomenon called the photoelectric effect. (It's interesting to note that Albert Einstein's Nobel Prize in 1921 was awarded for his work regarding the photoelectric effect and not for his theory of relativity.) It's important that we convert the photons into electrons because the energy associated with electrons can be easily measured as a voltage. Unfortunately, not all incoming photons result in an electron that can be counted. The percentage of photons that a detector converts into electrons is called the quantum efficiency.

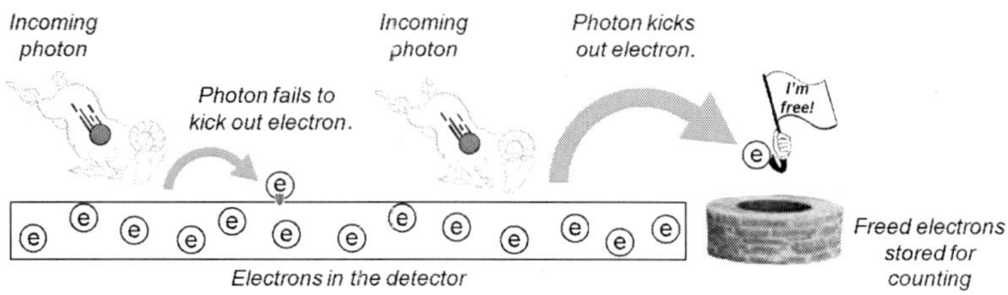

The photoelectric effect lets us count the light using electrons. Unfortunately, a photon does not always kick out an electron.

For a detector that is exposed to light, the bright light has more photons and will kick up more electrons in the detector than a dim light will. Electronic sensors store freed electrons in a well for each detector during the exposure time, so that they can be counted. When the exposure time ends, the sensor measures the energy of all the electrons in each detector well; this measurement indicates the number of electrons that were knocked free during the exposure.

The highest number of electrons that the detector well can hold is called the well depth. Care must be taken to limit the amount of light during the exposure time, so that the well does not overflow with electrons. If this happens, a flooding of electrons saturates the area with bright white pixels, an effect called pixel saturation. The amount of light hitting the detector

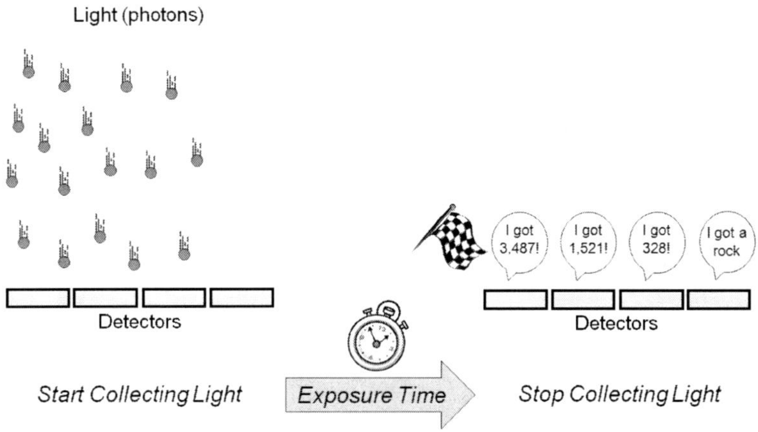

During exposure time, each detector effectively counts the amount of light it receives.

can be managed by using a shutter to control the time light can enter the camera, or by changing the aperture size (usually by changing the *f*/# setting on the camera). Keep in mind that limiting the amount of light too much leads to pictures that are too dark, so the shutter speed and aperture size must be just right to achieve bright pictures without saturation. Since most of us have no idea how to set the shutter speed and aperture size to find the right exposure time, many cameras have an autoexposure feature that figures all of that out for us.

When the exposure time ends, the number of electrons in each detector is measured to determine how many electrons filled the well; the higher the count, the higher the brightness of the light that hits that detector, corresponding to brighter light from the scene for that part of the image. The depth of the electrons in the well is measured with a ruler called a quantum step equivalence (QSE), which counts from zero for an empty well up to 255 for a full well. (The ruler measures the depth in increments corresponding to the number of available gray levels, so we will use a ruler that measures the depth in increments of 1/255 for 256 available gray levels.) A digital count value between zero and 255 is assigned to each detector based on the depth of electrons in its well. This digital count is the same number as that used to determine the gray level for a corresponding pixel in an image.

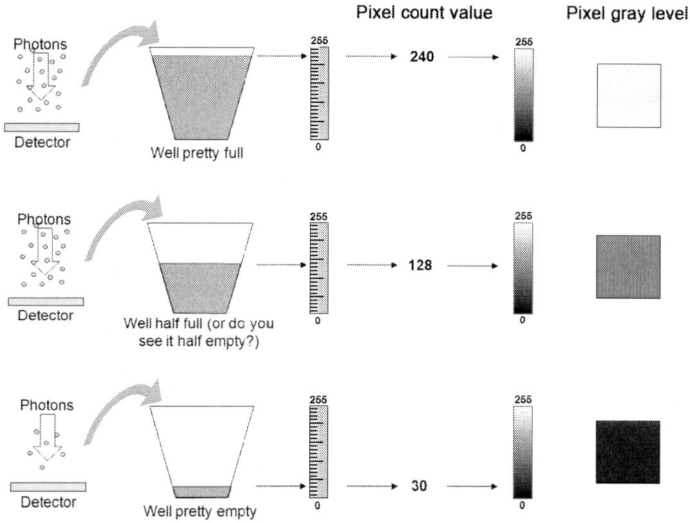

A certain amount of light hits the detector during exposure time and fills the well with electrons. The depth of the well is measured on a ruler that assigns a brightness number to the pixel.

If we flash back to the beginning of the book, you'll remember that we said a digital image is really just an array of numbers, where each number is a gray-level value between zero and 255 that tells the computer how bright to make each pixel on the display. Now we know that the gray level for each pixel is a result of measuring the

108	71	54	50
159	126	96	72
180	161	133	105
197	181	153	129

Numbers in a digital image...

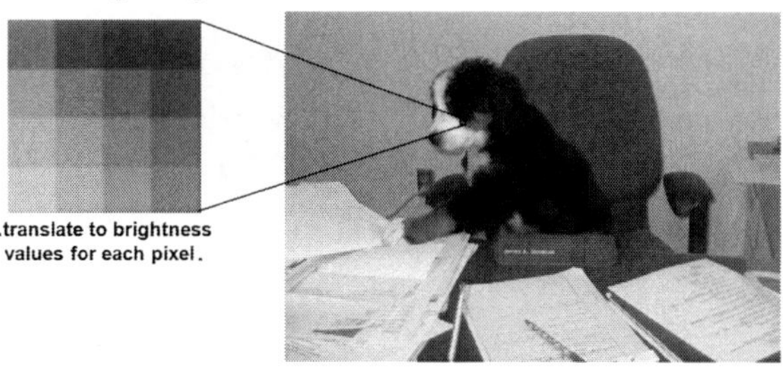

A digital image is simply a grid of numbers that tells a computer display how bright to make each pixel.

number of electrons that the light created at each detector. Using this array of numbers to tell the computer monitor how bright to make each pixel on the display results in a digital image!

6.3 Making Noise

The digital count for each pixel represents the brightness in a scene, but how well does it really represent the amount of light that illuminated the detector? Unfortunately, there are random uncertainties called image noise that prevent the number from

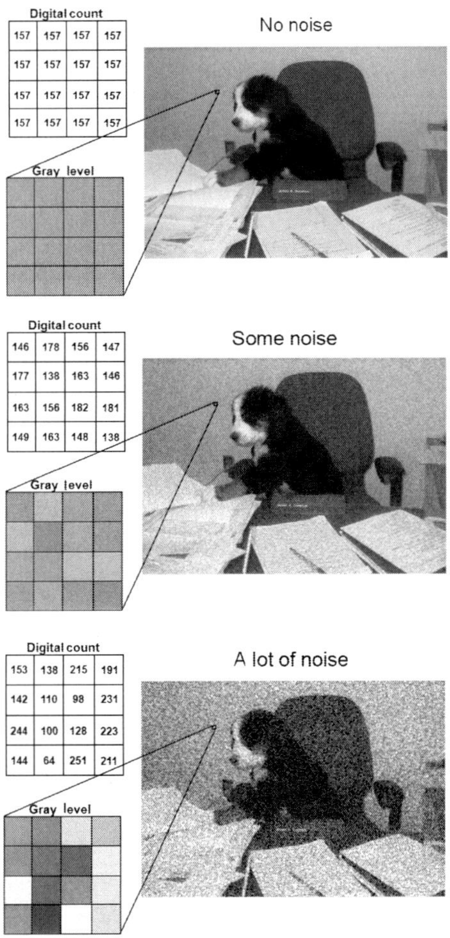

The random variation of each pixel's brightness is caused by image noise, which can be as annoying as a loud neighbor.

being spot on. Similar to audible noise, which is unwanted sound, image noise is an unwanted signal in our image. Image noise is one of the hardest image-quality degradations to deal with. As uncertainties worsen and values fluctuate more, a picture starts looking like somebody sprinkled salt and pepper all over it. The noise can become so great that we may not even be able to tell what we photographed.

Some primary noise sources are dark noise, shot noise, and quantization noise. Dark noise is a fluctuation of count values, even when there is no light, and has many contributors, including electrons generated from the warmth of the detector. Shot noise, also called photon noise, is caused by a fluctuation in the arrival rate of photons at the detector. Quantization noise occurs because the ruler to measure the electrons in a well can only measure depth to an accuracy of 1/255, so a range of electrons is given the same digital count value.

The important measure for camera designers trying to manage noise is how the strength of the light signal compares to the fluctuations caused by noise. This comparison is called the signal-to-noise ratio, or SNR. Unfortunately, there are many different ways to calculate the SNR, so saying that a picture has an SNR of 55 does not really mean much unless we are told how the SNR is being calculated.

6.4 Sampling and Psychedelic Pinstripes

As an astute reader, you noticed that each detector only reports one count value for all of the light that fell on its surface. This limits a digital camera's ability to create an image that has the same detail that we see in the scene. First, every scene detail that is imaged across the surface of the detector is mixed and added together. Second, the detectors produce one number for each detector, but the detectors are spaced across the sensor, with the detector-to-detector distance called the detector sampling pitch. By putting these two effects together, we can think of the scene as first being blurred from the mixing of the light across each detector, and then as this blurred scene being sampled at intervals corresponding to the distance between the detectors (detector sampling pitch).

The sampling effect on an image is very important to understand because it can dramatically change the way objects in a scene appear in an image. It is also critical because this sampling places a fundamental limit on the details that we can see in the image. Once again, it is easier to understand the effect sampling has on an image if we look at a wave pattern in the scene. To see the wave pattern, there needs to be enough detectors across the wave to recreate it in the image. At a minimum, there needs to be a detector for every bright band and every dark band. Sampling at this limit is called Nyquist sampling, named after Harry Nyquist, who figured out the limit while working at Bell Labs to determine the limiting number of pulses that could be sent across a phone line. If there are not enough detectors across the wave pattern, not

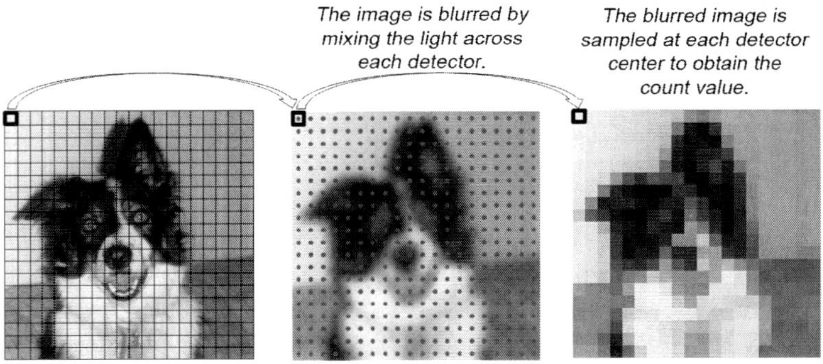

Blurring and sampling of a scene by the detector limits the detail that we can see in an image.

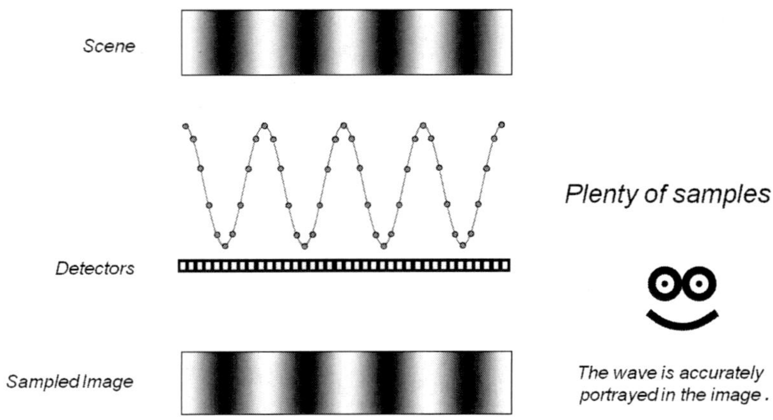

Many detectors across a wave pattern will faithfully reproduce the wave pattern in an image.

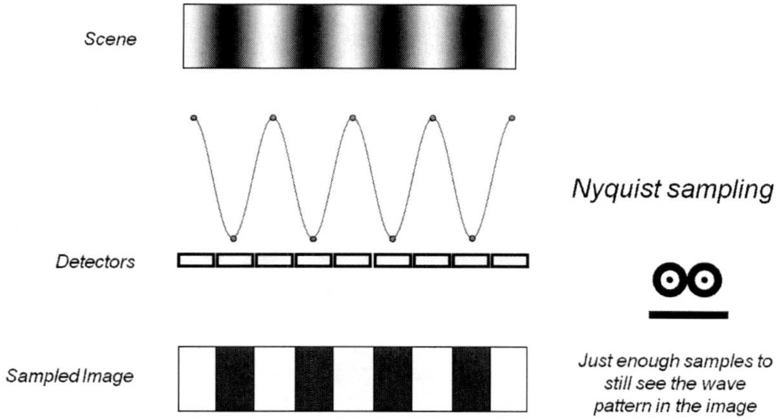

Nyquist sampling occurs when there are just enough detectors across a wave pattern to image each bright and dark band in the wave.

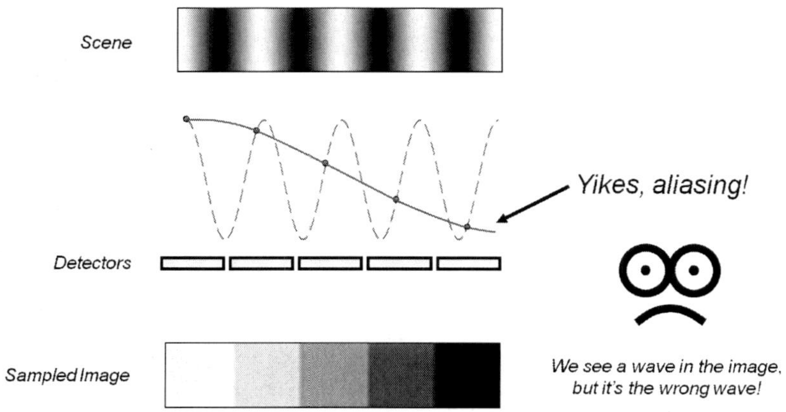

When there are not enough detectors across a wave pattern to image each bright and dark band in the wave, an alias wave pattern shows up in the image that is not in the actual scene.

only is the wave pattern not seen in the image, but a new *alias* wave pattern is created.

Just like a person who uses an alias for his name, aliasing in an image can be problematic because we see something in the image that isn't true to the actual scene. This aliasing effect is seen on TV when a person wearing a pinstripe suit appears and we see crazy wave patterns shimmer across the suit as he moves across the screen. The aliasing on the TV occurs when the pixels on the TV screen are spaced farther apart than the distance between the pinstripes on the suit.

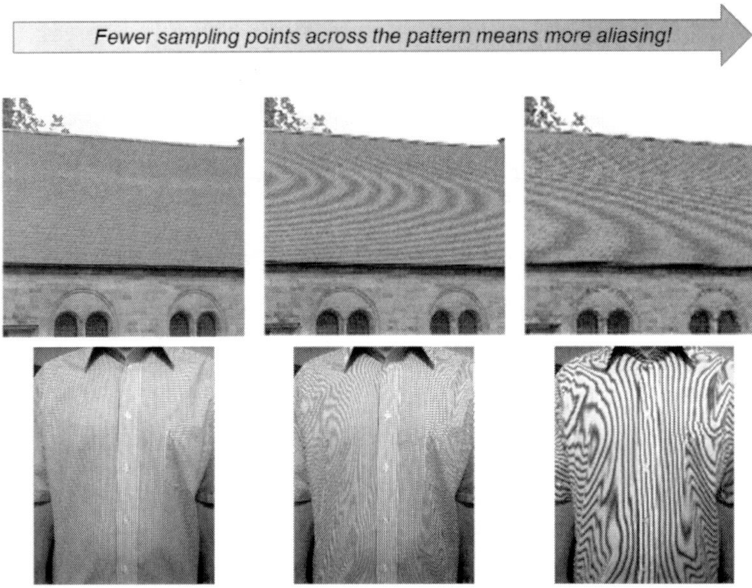

Alias wave patterns grow more noticeable as the number of sample points across the pattern decreases.

Chapter 7
Camera Considerations

7.1 Making Sure the Left Hand Knows What the Right Hand Is Doing

It would be nice if we could simply take the best light source, the best optics, and the best sensor and put them all together to get the best camera. Unfortunately, that would be like taking the best player for each position from a pool of professional football teams and assuming that you have automatically created the next Super Bowl championship team. As coaches know, the interaction between players is a critical part of a winning team, and the same is true with digital camera design. The optics and

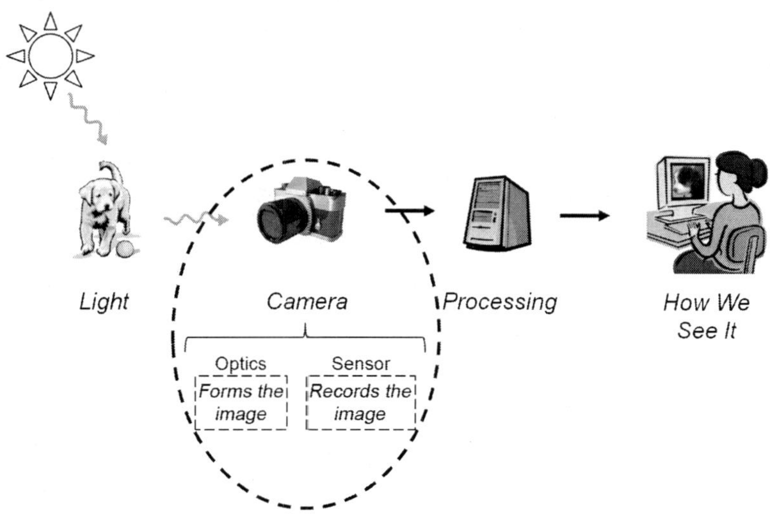

Designing the camera as a whole requires considering how the components will work together with light, motion, and other aspects of imaging.

sensor need to be designed together to create the best camera for the intended purpose, such as family photos or stellar astronomy, within given constraints, such as size and cost. This approach to designing the system as a whole is called *systems engineering*, and it considers how the individual parts work together to create the best image with the existing light and motion present when snapping the picture. First, we will look at the effects of motion blur on image quality, then, we will see how this influences the design of the optics and sensor.

7.2 Taking Pictures in a World that Won't Stand Still

We can all relate to that time when we quickly snapped a photo, and after our elation of "getting the shot" settles down, we notice that the photo is blurred from motion. If only we had used a tripod to hold it steady, but then we probably would have missed the shot. Motion blur often occurs from an unsteady hand, and other times it is from the object moving in the scene, such as a rambunctious puppy that just won't sit still for the picture.

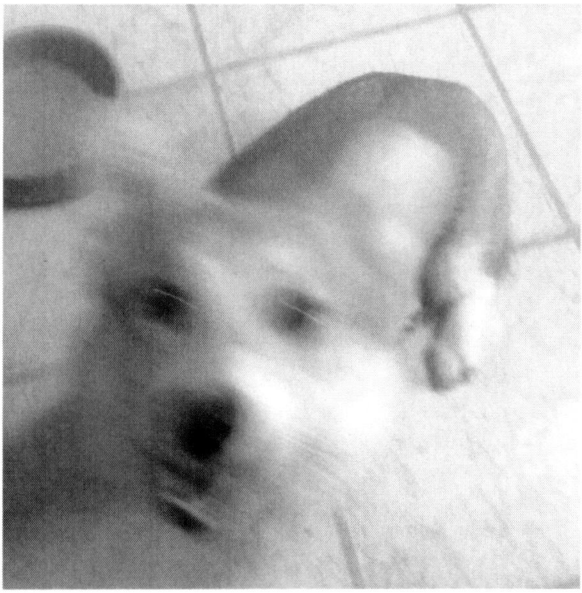

Capturing that special moment in a picture is difficult when the subject doesn't cooperate during the camera's exposure time. It can even make a cute puppy look rabid!

Motion blur is generally described as either jitter or a smear. Jitter blurring occurs when the camera moves in a relatively fast, random motion in all directions (a camera held in a shaky hand), and the image blurs equally in all directions. We've all experienced the difficulty of trying to hold a camera steady while taking a picture of lights at nighttime, only to see the lights blurred into blobs in the final picture due to our unsteady hand. The severity of the blur depends on the magnitude of the shaking during the exposure, so we try to steady the camera as much as possible.

Smear is a blur that occurs in a single direction, specifically when there is a straight-line motion between the camera's sensor and the object being imaged. The classic example of motion smear is a speeding car that is blurred when we snap the picture as it whizzes by. When the camera moves in a straight line with respect to the scene, the entire image will exhibit motion smear. It's important to note that motion smear occurs in only one direction, so it can have a different effect on an image, depending on the direction of the smear.

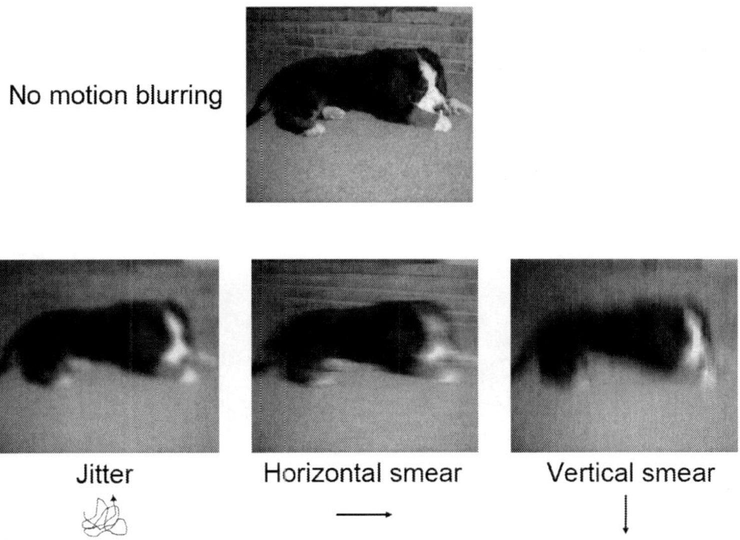

Blurring caused by camera motion has different effects, depending on the direction of the motion.

Another motion effect worth noting is an oscillation, which can be seen in images from cameras that scan a scene to create the image, rather than taking a single snapshot. A scanning approach is used when we need more pixels across an image than the largest available rectangular detector array can provide. The detectors are placed in a long line instead of a rectangular array, then, they scan the scene to create an image one row of pixels at a time. This approach is used for copiers, scanners, and even imaging satellites. If the camera moves back and forth while the scene is scanned, the straight lines will appear to have a wiggle. Oscillation in the image will have different characteristics, depending on the direction of the oscillation.

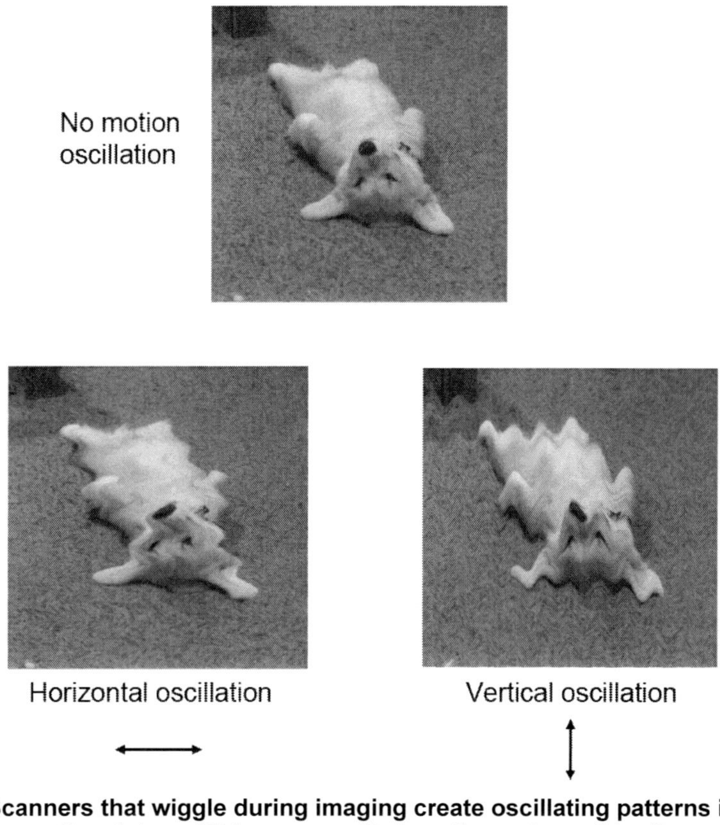

Scanners that wiggle during imaging create oscillating patterns in the picture. (This puppy will sleep through anything!)

So, what can we do about minimizing motion blur in an image? For one thing, we can lock down the camera so that it doesn't move while we take the picture. We do this all the time when we use a tripod to hold the camera, but it's not always convenient to whip out a tripod every time we want to take a picture. Many cameras have an image-stabilization feature that attempts to counteract any camera motions. The problem of stabilizing the camera is even more challenging when it is on a fast-moving airplane or satellite.

Another way to minimize motion blurring is to reduce the exposure time so that there isn't much time for things to move while the photo is being taken. This, however, cuts down on the amount of light used to make the picture and can produce unacceptably dark pictures, especially if there isn't much light to begin with. To compensate for the loss of light, we can build a bigger camera aperture to collect light, or use a flash to add more light to the scene. The autoexposure feature on many cameras

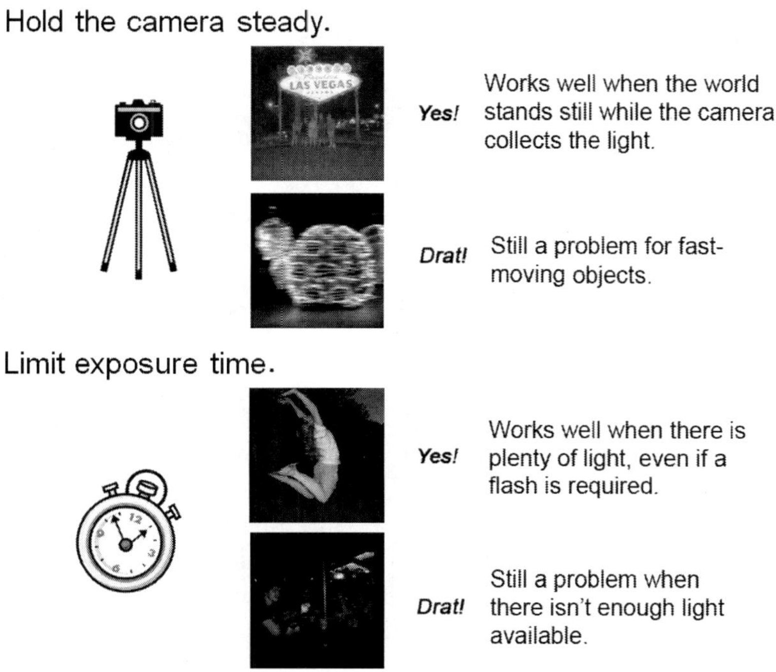

Motion blur can be reduced by holding the camera steady and limiting the exposure time. Ideally, we would like to do both.

helps by measuring the light available to determine the shortest exposure time possible that will still produce a bright image.

7.3 Seeing the Detail: Resolution and Camera Size

The smallest detail that can be seen in a digital image is measured by the camera's resolution limit, a measure of the minimum distance between two objects in a scene that will still look like two separate objects in an image. When we are asked to read an eye chart, the optometrist is determining how well our eyes can image the details. One example is the tumbling-E chart, where the optometrist asks us which way the letter E is pointing to determine how close the bars on the letter E can get before we can no longer see them as separate bars.

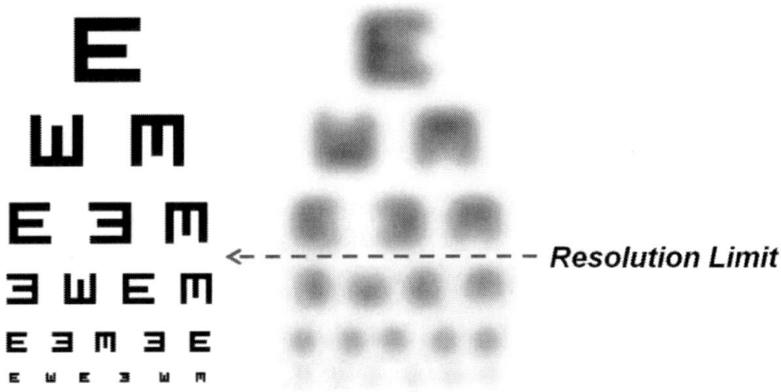

The resolution limit tells us how close two objects can be in a scene and still look like two separate objects in an image.

As we magnify an image on our computer display, we notice that there is a point at which we are no longer seeing more detail or information in the image. When this happens, you've just determined the resolution limit of that image. What limits the detail that can be captured in an image? Many factors contribute to the final resolution limit, such as available light, but the camera as a whole has a fundamental limit on the resolution that can be captured in an image, and this limit is determined by the attributes of the optics and sensor.

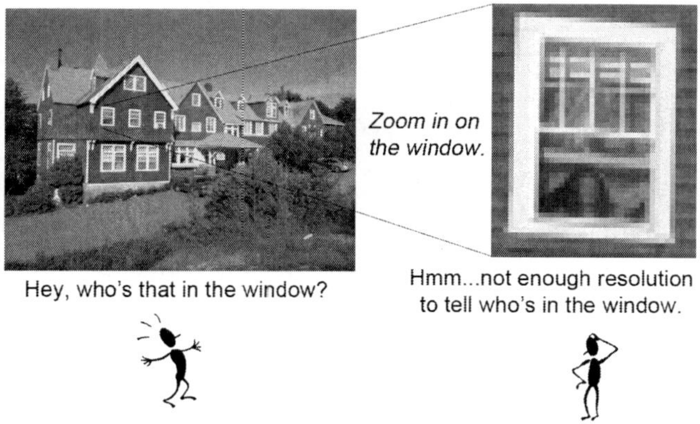

The detail that we can see in the image is limited by the camera's design, even after zooming in.

As we saw earlier, the camera is usually thought of as two individual parts, the optics and sensor. All cameras, from cell phones to the cameras that take pictures from space, have an optics component to collect light and a sensor component to record it. We also saw earlier that the optics and digital sensor each have a fundamental limit to the detail that each can capture in an image. The optics is limited by the spreading of light, called diffraction, as it passes through the camera aperture, and the detector is limited by the distance from one detector to the next, called the detector sample pitch.

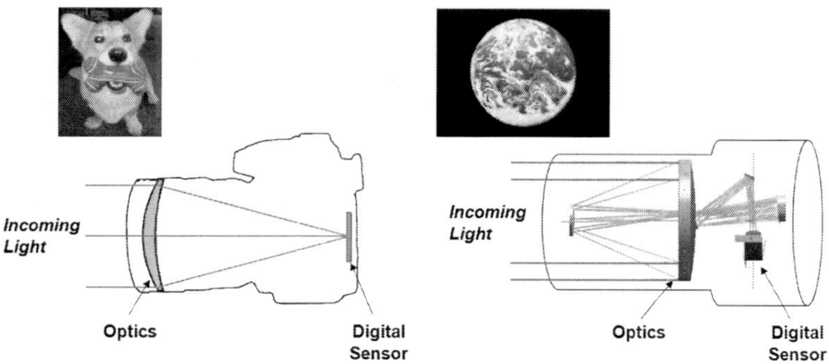

Hand-held digital camera ***Satellite digital camera***

All digital cameras have optics to collect light and a sensor to record it, but these both limit the details that we can see in the image.

What can we do to improve the resolution in an image? From the perspective of the optics, resolution becomes better if we use light with a shorter wavelength (for example, blue instead of red), if we use a bigger camera aperture, or if we move the camera closer to the object. From the perspective of the sensor, resolution becomes better if we reduce the distance between the individual detectors by using smaller detectors, if we use a longer focal length lens, or if we move the camera closer to the object. Notice that the one thing that we can do to improve resolution from both the optics *and* sensor perspective is to move closer to the object. We do this all the time with our personal camera when we move the camera closer to get the shot with more detail.

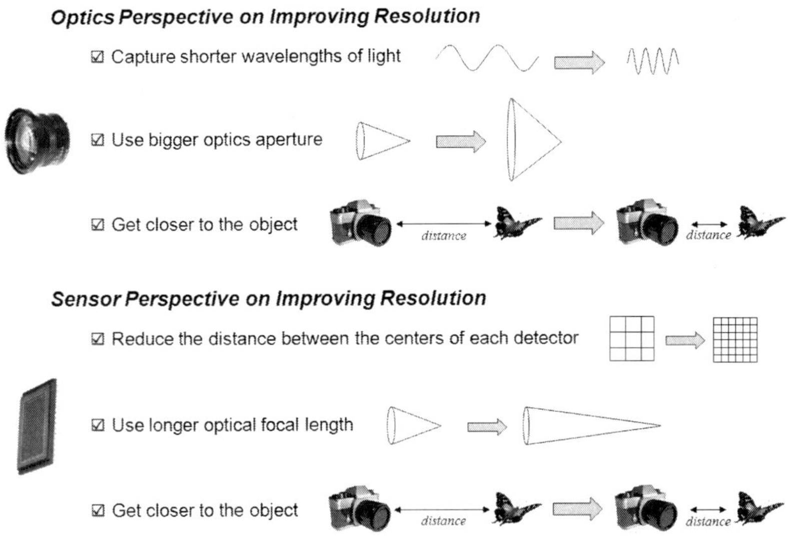

The optics and sensor each limit image resolution. Moving the camera closer to the object improves both limitations.

Astronomers cannot move closer to the stars, so they build bigger telescopes with longer focal lengths. Note that the earliest telescopes had apertures as large as lens makers could make them, and the lengths were very long to accommodate the long focal length. Telescopes today use large mirrors that can be built lighter than the earlier heavy glass lenses, and multiple mirrors are typically used to *fold* the optics, resulting in a shorter telescope. A long focal length magnifies the image of the stars but also magnifies any motion blurring that occurs. This is why telescopes need to be held very steady while they collect the precious little light that arrives from distant stars and galaxies. Another significant problem is the turbulence in the air that causes the stars to twinkle but also causes them to blur in the image. Astronomers like to place their telescopes at high elevations where there is less air to cause this blurring. The Hubble Space Telescope is in the ideal location, far above any atmosphere.

Astronomers have the right idea to build cameras that will produce the best image quality. But wait a minute, a bigger camera aperture means a bigger and heavier camera, which isn't appealing to people who would like to carry a camera in their pocket. Size and weight is also a big problem when considering launching a camera into space because a bigger and heavier camera means a bigger and more expensive rocket to get it up there. Size and weight are important considerations when designing a camera to meet the needs of the people who will be

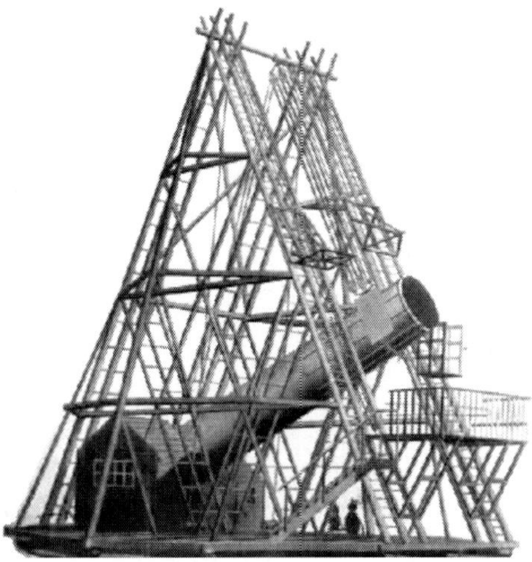

Herschel's 40-foot telescope in 1789

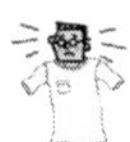

Not exactly "pocket size"

Early astronomers, such as Sir William Herschel who discovered Uranus, knew that bigger telescopes would give them better resolution. Current camera designers may want to give up resolution for other considerations, such as the ability to put the camera in a pocket.

using it. Another consideration that we will discuss next is balancing the resolution limits of the optics and detector to give the best image quality for the size and motion of the camera.

7.4 The Story of Q: Designing the Optics and Sensor in Harmony

Another important consideration is that a bigger aperture might not buy you any more resolution in the image if it is already limited by sensor resolution. If a camera is designed so that the optics resolution is driving the overall image resolution, the image will look blurry. On the other hand, if a camera is designed so that the sensor resolution is driving the overall image resolution, the image will look blocky with those aliasing patterns. The optics and sensor resolution limitations need to be balanced to offer the best image quality for the intended use of the images. What is the best camera design that balances the optics resolution limit and the sensor resolution limit? This question has been debated for decades because the design that mathematically balances these two limits is not the best design for all cameras. Let's take a closer look at the question to see what the debate is all about.

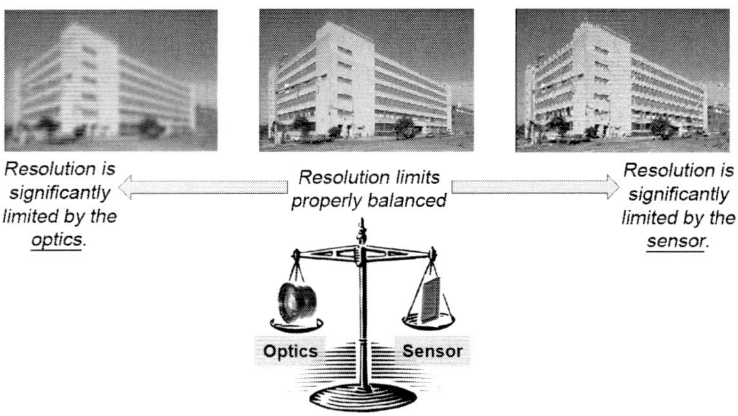

Designing a camera to balance the resolution limits between the optics and sensor is critical for maximizing image quality.

Camera designers derived a parameter to measure the balance between the optics and sensor resolution that has historically been called $\lambda FN/p$ (pronounced lam-duh-eff-enn-over-pee), but even for scientists and engineers, that is a mouthful, so now it is simply called Q. The letter Q has no specific meaning; it was simply chosen because we were looking for a letter in the alphabet that wasn't already being used for digital camera designs, and it's quicker to say than $\lambda FN/p$.

In a nutshell, the parameter Q tells us how much better the sensor resolution limit is compared to the optics resolution limit. A high-Q camera has a sensor resolution limit that is better than the optics, but keep in mind that the resolution of the *camera* is limited by the worst part; hence, the camera resolution for a high Q is limited by the optics. Conversely, a low-Q camera has better a better optics resolution limit, meaning that the sensor resolution is limiting the camera performance.

The value of Q for camera design can be altered by changing those parts of a camera that modify either the optics or sensor resolution. Earlier we showed that these were the wavelength (color) of the light, the size of the optics aperture, the focal length of the optics, and the distance between the neighboring detectors in the digital sensor. Changing the distance between the camera and the object we're taking a picture of changes the resolution but doesn't change the value of Q because both the optics and sensor resolution are changed by the same amount.

It seems reasonable then to design a camera that has equivalent optics and sensor resolution limits, so that we aren't wasting money by overbuilding one part to capture details that the other part can't handle. Mathematically, the two resolution limits are equal when Q has a value of 2. You might be wondering why the value of Q is not unity when the two resolution limits are the same. The simple reason is that Q was historically defined using the detector sampling distance, which is different from the sampling resolution limit by a factor of 2. (Remember, Nyquist sampling required two detector samples, one to see the bright band and other to see the dark band, for one period of a wave.) A camera with a Q higher than 2 will have a sensor capable of capturing more details than the optics can deliver. A camera with a Q less than 2 will have optics that captures details thrown away by the sensor. So, it seems as though digital cameras should be designed so that Q has a value of 2. But here's the kicker: most cameras are built so that Q has a value *less than* 2! Have the camera designers lost their minds? On the contrary, a good camera designer looks at optimizing the overall quality and usefulness of the pictures, and it's the simple mathematical analysis that misses the point.

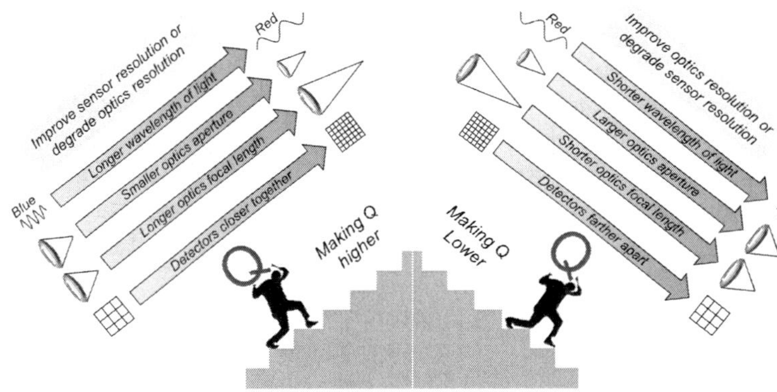

The Q of the camera can be changed by changing the parts of a camera that alter the relationship between the sensor and optics resolution limits.

The confusion comes primarily from the fact that Q only compares the resolution limits between the optics and sensor, and does not consider the overall image quality. Many other camera design parameters need to be considered, such as the lighting condition and stability of the camera. An important design consideration is the robustness of the camera to take good-quality pictures under poor lighting or with motion present. It turns out that lower-Q cameras make brighter and sharper images and are less susceptible to motion blurring, so the camera designers like to push Q a little bit below a value of 2 to get some additional benefits. The Q is generally lowered to a value between 0.5 and 1.5 by oversizing the optics with respect to sensor resolution. The larger aperture also collects more light, allowing for faster exposure times that further reduce motion blur. As mentioned earlier, the size of the camera is an important consideration that cannot be ignored, so designing a larger aperture to reduce Q might not be the best overall solution in certain cases. The camera designer must therefore weigh all of the options available in order to offer a camera that is designed to best meet the needs of the intended user.

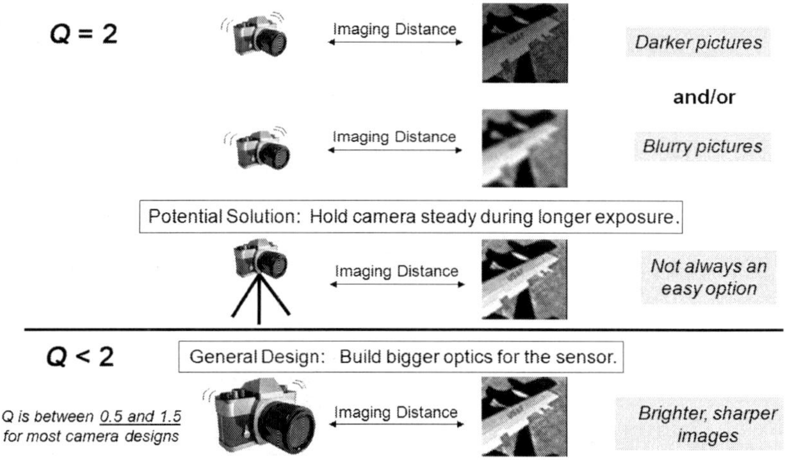

Camera designs with Q value below 2 can produce images that are brighter, sharper, and less susceptible to motion blurring, but they typically use a larger camera lens, which might not be practical.

Chapter 8

The Magic of Computers: Making a Picture Better

8.1 The Numbers Game

The next link in the imaging chain is the point where the camera's sensor turns an image into an array of numbers, typically between 0 and 255. The sensor divides the image into picture elements, or pixels, and assigns a number to represent the brightness of the image at each pixel. (A color camera creates three arrays of numbers, corresponding to the brightness of the red, green, and blue pixels.) Mathematicians love this, because they can now play with the numbers using various equations to alter the look of an image, or to extract information from the image data. An image can be processed using various

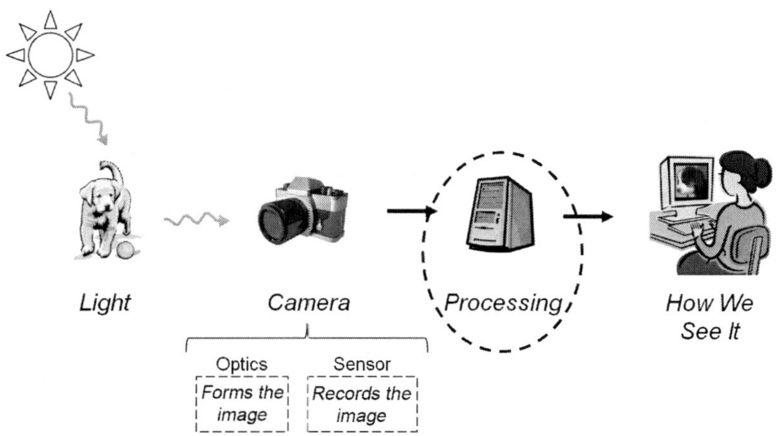

Processing can make a picture look better by changing the numbers in the digital image.

mathematical formulas to change the numbers in an attempt to improve the way the image looks. Of course, beauty is in the eye of the beholder, so people do not necessarily agree on the exact process that results in the best image, but that is the subject of the next chapter!

We need to be careful when we change the numbers because our eyes are very sensitive to changes that do not look normal. If the numbers are not changed in a smart way, the look of the image can easily become worse than the original. Sometimes it's fun to change the numbers to intentionally warp a normal-looking image, either for an artistic impression or simply for a chuckle.

Let's look at a few quick examples of ways that we can change the numbers. If we divide each number by two, the image will lose half of its brightness. If we move the numbers twenty pixels to the right, the image will appear to shift to the right. We can also move the numbers away from the center of the array to warp the image, but this requires a method for filling in the blank areas that will be discussed later in this chapter.

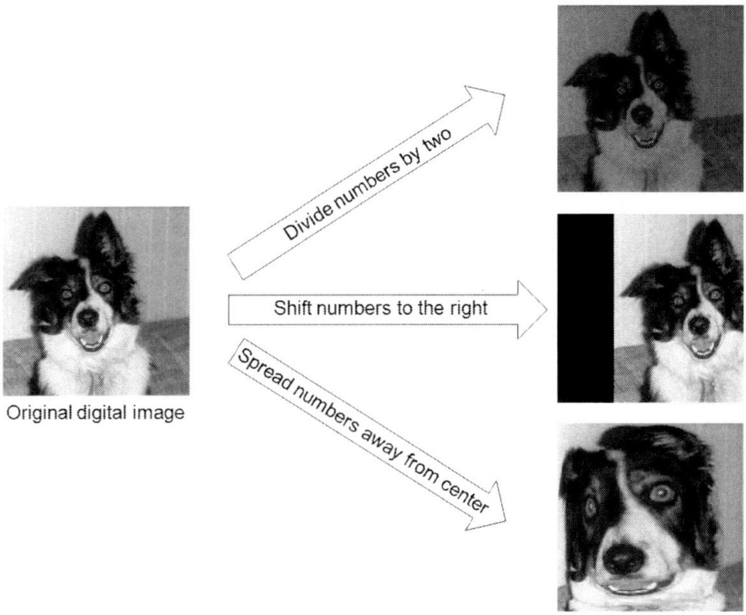

Image processing uses a set of instructions to change the gray-level values for each pixel, sometimes in a fun way.

We will look at two processes that have been developed to enhance the quality of the image. The first, called contrast enhancements, improves the brightness, contrast, and tone of an image. The second, called spatial filtering, reduces the blur and improves the clarity of an image. These two processing steps used together improve the brightness and clarity of digital images captured by a camera. Many times these steps are actually part of camera-sensor processing, so the enhancements take place before we ever look at the picture.

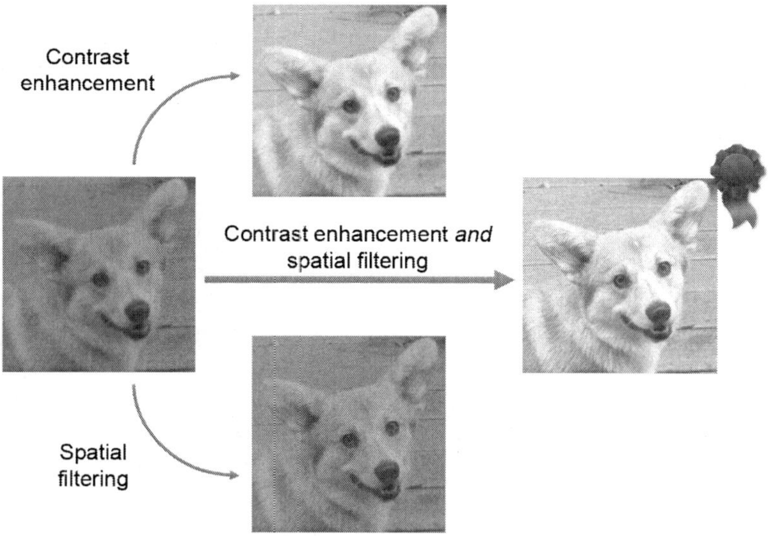

A combination of contrast enhancement and spatial filtering provides brighter pictures with sharper details.

8.2 Taking Away the Drab

We've all taken pictures that were too dark and wished we could brighten them up to see the details better. Other times we've taken pictures that look very gray and washed out and wished we could improve the contrast to make the grays darker and the whites brighter. Now your wish can be granted! We can change the numbers in a digital image to adjust the brightness and contrast to our liking. Remember, low gray-level values produce dark pixels and high gray-level values produce bright pixels.

To help us understand the contrast of an image, we need to see how the gray levels stack up by looking at the gray-level histogram of an image. A gray-level histogram is created by counting the number of times that each gray level occurs in an image, and then plotting the results in a bar chart. Think of stacking poker chips into 256 separate stacks lined up in a row, with each stack labeled sequentially with a number from 0 to 255. We then look at the gray level for each pixel and add a chip to the stack that has the same number as the gray level. When we've added a chip for every pixel in the image, we can stand back and look at the profile of stacks to see how the gray levels are distributed from 0 (black) to 255 (white). By simply looking at the gray-level histogram of an image, we can tell a great deal about the brightness and contrast of the image. Dark images have most gray levels piled up in the low end, while bright images have most gray levels piled up in the high end. A low-contrast image has a skinny gray-level histogram because there isn't a large range of brightness values; this causes the image to look drab and hazy. A high-contrast image has a wide gray-level histogram because there is a large difference in brightness between the darkest parts of the image and the brightest parts.

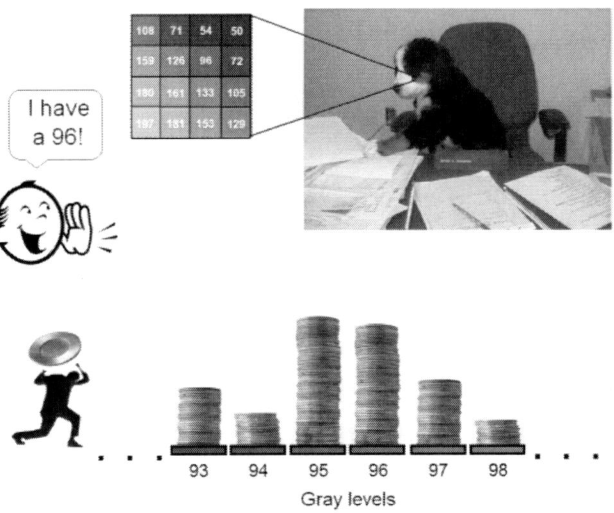

A gray-level histogram is built similar to stacking chips into 256 stacks, with one chip added to a stack every time a pixel has the corresponding gray level.

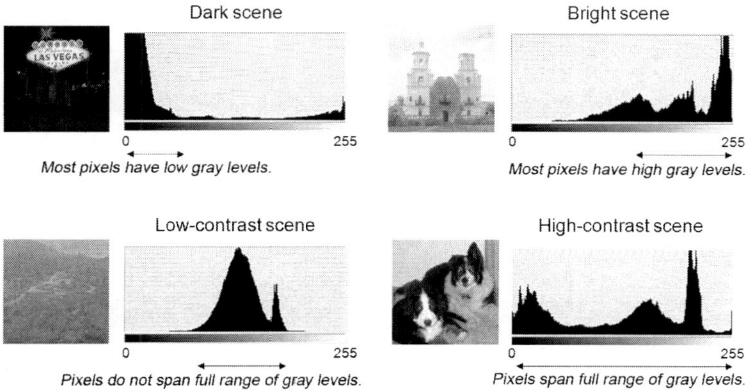

Gray-level histograms for images with different brightness and contrast.

What can we do with a gray-level histogram? The histogram helps us determine how to change the gray levels at each pixel, such that the resulting image has the desired brightness and contrast. For example, we can create a negative image by changing the gray levels so that the histogram is flipped because this reverses the blacks and whites in the image. The histogram becomes flipped if we simply subtract the gray level at each pixel from 255.

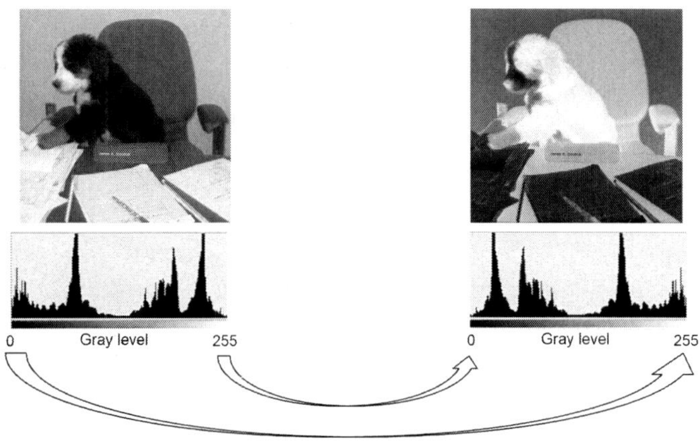

A negative image has a gray-level histogram that is a mirror image of the original histogram because what was white is now black and what was black is now white.

When we have an image that looks gray and dreary, we can improve the contrast by making the darks darker and the whites whiter. The simplest way to do this is to reassign gray levels to each pixel that effectively stretch the gray-level histogram wider to span more of the available gray levels in the 0–255 range. Be careful; stretching a histogram creates gaps between the resulting gray levels, causing jumps in the brightness values in an image that can create a contouring effect.

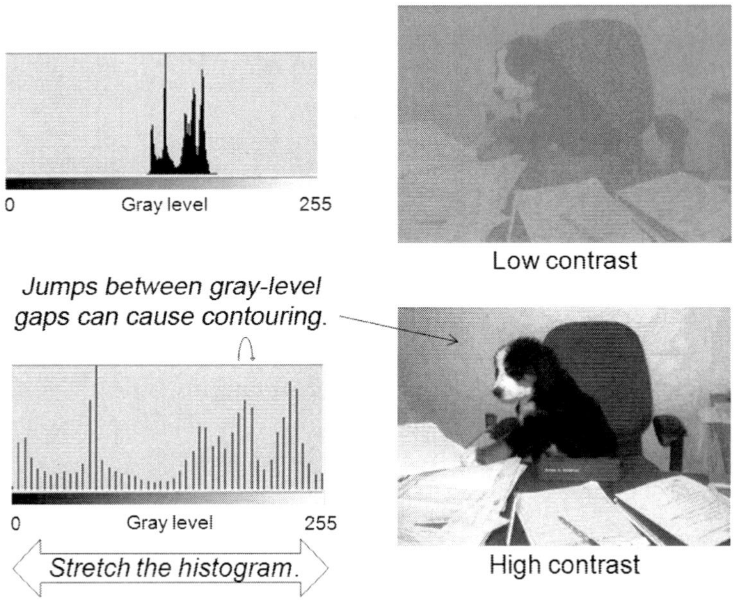

Stretching a low-contrast image can create contours in the image that are not in the actual scene.

Contrast can be raised by stretching a histogram even more, but there is a limit. If we continue to stretch the contrast, gray levels stretched below zero will all be assigned a gray level of zero because the computer cannot interpret gray levels below zero. Likewise, values stretched above 255 will be assigned a gray level of 255. This *clipping* of the histogram makes the dark grays black and the light grays white, which causes details in those areas to be saturated and lost. Normally, a little bit of clipping is a good thing, but we can't get carried away or we'll end up losing precious details in the image.

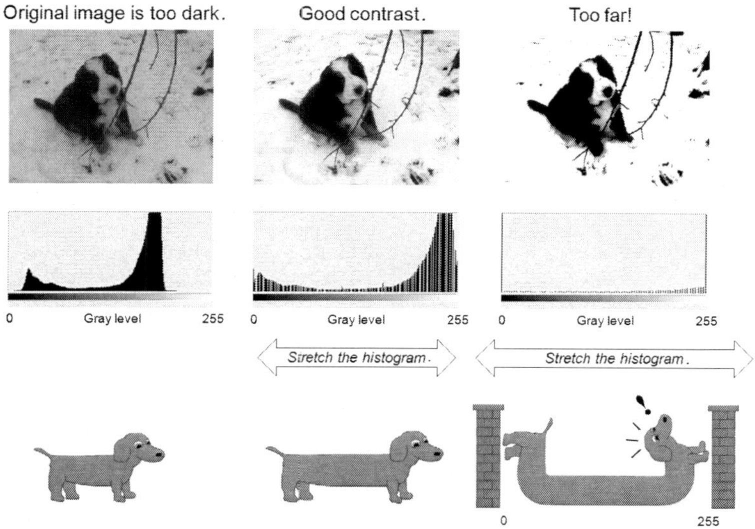

Stretching the contrast too much saturates the dark and bright regions and causes a loss of detail.
(Dog illustrations courtesy of Katie Fiete.)

So far we've only talked about stretching a histogram evenly across all gray levels. But a histogram can be stretched unevenly to enhance dark regions more than bright regions or vice versa. This processing is generally referred to as tonal enhancement and is a great way to pull out details from dark shadows or dark details in bright regions.

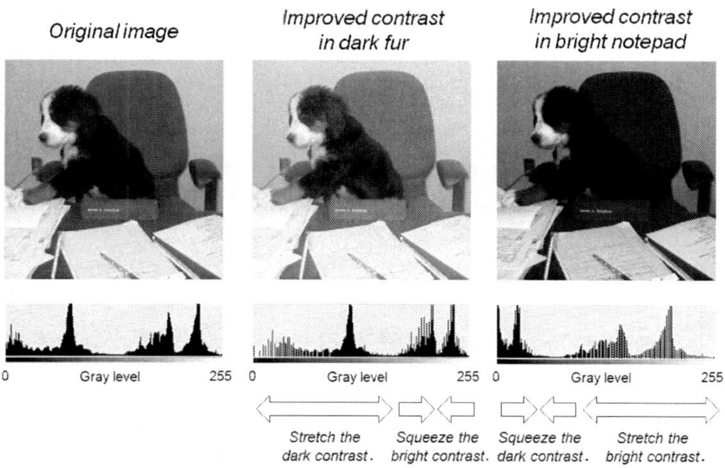

The contrast of dark regions can be enhanced at the expense of the bright regions or vice versa.

8.3 Eyeglasses for a Digital Image

Now that we have seen how image processing can improve the brightness and contrast, let's look at what we can do about the blurriness of an image. Earlier we talked about thinking of the camera as a linear shift-invariant system because this makes the mathematics for understanding how a camera works much simpler. We can take advantage of this viewpoint to change the relationship of the numbers between neighboring pixels in a digital image through a process called spatial filtering. The mathematical operations are called filters because we can use them to filter out certain aspects of an image that we may not wish to keep. We cannot use this process to filter out an ex-boyfriend or an ex-girlfriend, but we can filter out some of the characteristics that a camera places on an image that we are not happy about, such as blurring.

Spatial filtering changes the gray level of a pixel based on the gray levels of its neighbors. As an example, we could create a new image by instructing the computer to make a new pixel value the difference between the current value and its neighbor to the left. If we process the whole image with this instruction (called a derivative filter), we achieve an image that highlights the vertical edges in the image. If we add together all of the neighboring gray levels, we are mixing all of the local information, resulting in a blurred image. (Remember that gray levels can only be between 0 and 255 for an 8-bit image, so frequently we must rescale the numbers so that they fit within the range of 0 to 255.)

Now we know that a filter that adds the neighboring gray levels together blurs an image. Is there a filter that unscrambles the neighbors to take away blur from an image? Though it seems counterintuitive, the answer is yes, but only for a limited amount of blurring. By understanding how the imaging chain blurs an image, we can create a filter that removes some of the blurring. On many image and photo-editing software packages, such as Adobe Photoshop®, the spatial-filtering application to reduce blurring in a photo is labeled *sharpen* under the filtering options because it produces an image with steeper and sharper edge profiles.

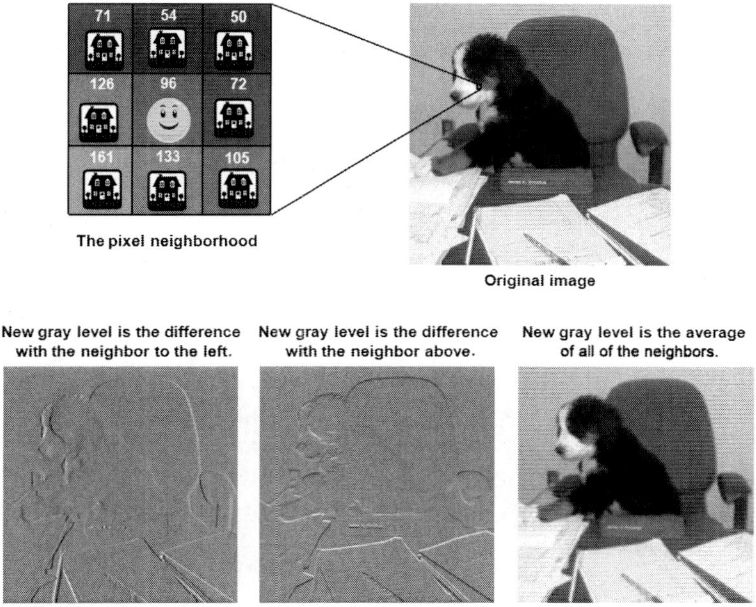

Spatial filtering creates a new image by creating relationships between a pixel and its neighbors.

Of course there is no free lunch here. There are three problems that we encounter when we design a filter to remove the entire blur. The first is that much of a scene's detail is blurred away because the optics and sensor can only capture scene detail up to a point. Removing the entire blur requires us to put back the detail that was outside that resolution limit, and unfortunately that information is now dust in the wind. The second problem is that a *ringing* artifact of black and white bands starts to show up along the edges as we try to remove more of the blur. The third problem is that the filter that removes the blur also has a nasty habit of enhancing details in the image that we don't like, such as image noise and aliasing. Filters are designed to reduce blur, but they cannot take it all away. These sharpening filters are usually tweaked to provide the right balance between sharpening and enhancement of unwanted artifacts.

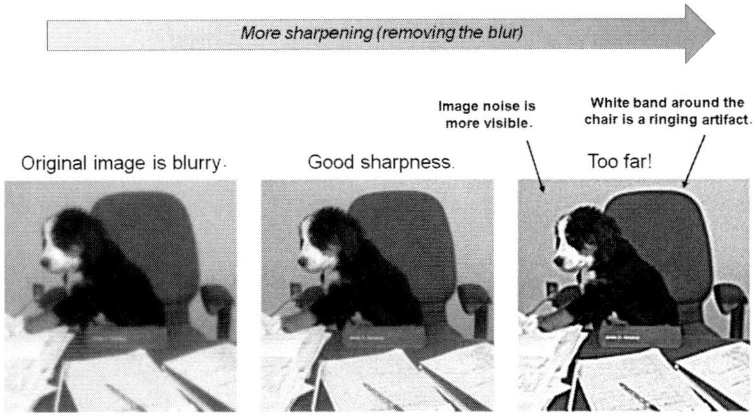

Sharpening filters can remove much of the blur, but removing too much produces unwanted features in an image.

8.4 Changing Size and Filling in Details: Reality Versus Hollywood

A digital picture from our camera usually has more pixels than our computer display has available, so there is not a pixel on the display for every pixel available in the image. For example, a ten-megapixel camera has ten million pixels, but an HD display only has about two million pixels. The size of the image is reduced to view the entire image on the display, thus throwing away pixels that the camera captured. But we know that our image has additional details that we cannot see due to this downscaling, so we zoom in to make the features bigger on the smaller display. If we keep zooming in, we notice at some point that the features keep getting bigger, but there is no more detail to see. When this happens, there are no more pixels in the image to provide additional detail, and there are now more pixels on the display than were in our original image. The computer added new pixels to the image to fill in the display pixels, so that we don't see holes in the image. How does the computer fill in the holes? The process of filling in the holes is called interpolation.

Interpolation between pixels is actually necessary any time the scale of an image is changed and the pixels in the scaled image do not line up exactly with pixels in the original image. The interpolation method uses an interpolator, which creates a smooth transition between the gray levels of pixels in a scaled image based on the gray levels of pixels in the original image. Image scientists have developed many different ways to fill in the gaps, each with their own advantages and disadvantages.

The simplest way to fill in the holes is to simply repeat the pixel gray-level value of the nearest pixel. This method is called nearest-neighbor interpolation and is very fast on a computer because it simply copies a number. However, this method leads to an image that looks blocky. Another method, called bilinear interpolation, finds the average of gray-level values around a pixel. This method leads to a smoother transition between existing pixels but adds a blurring effect to the image. Slightly more complicated methods, such as the bicubic interpolator, balance the smooth transition with

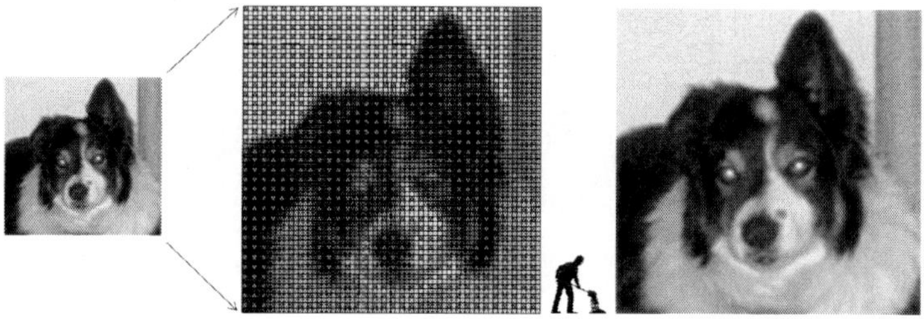

Magnify 2× by putting a hole between each pixel. Interpolation fills the holes.

An image can be magnified to twice its size by placing holes between the image pixels. Then an interpolator fills in the blanks to produce a nice-looking image.

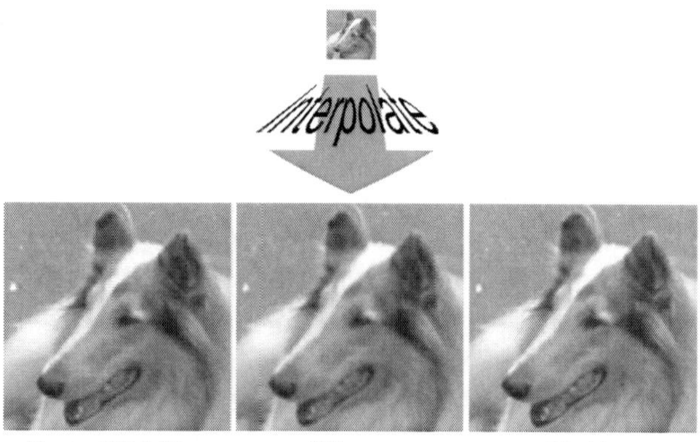

Nearest Neighbor Bilinear Bicubic

Different interpolators, each with their own advantages and disadvantages, are designed to assign gray levels to the new pixels that are created when an image size changes.

less blurring, but they require more computations, which is usually alright if you don't have a slow computer.

Let's take a moment to talk about the magical way that people on TV shows, such as crime scene investigators, are able to continuously zoom and enhance an image. My personal favorite is the TV show where they zoom in on an image from a surveillance camera to see the reflection of the killer in the victim's eye. Catching criminals is easy and fun with the magic of image processing! Yes, this is one of those "Oh, come on!" moments, but I'm always surprised at the number of people who think that this is a common technique used by crime solvers. Understanding the imaging chain allows us to realize the limitations of a digital camera and keeps us from being fooled by these Hollywood stunts. But is this really science fiction, or could it be possible?

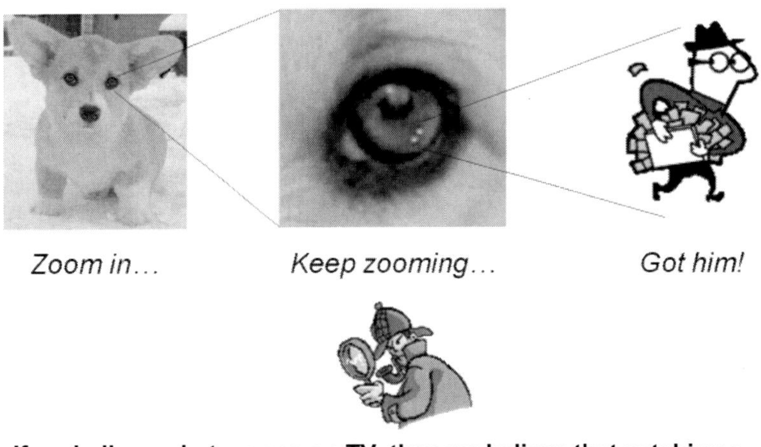

If we believe what we see on TV, then we believe that catching a criminal is as easy as catching his reflection in the eye of a witness.

We saw in the previous chapter that the resolution and detail that can be seen in an image is fundamentally limited by a camera's optics and sensor. We can work backward from this understanding to see if a camera is actually capable of providing an image with the detail necessary to identify a criminal from a reflection on the eye. The truth is that no laws of physics are being broken here, but building a video surveillance camera with this capability poses some serious technical and practical challenges. The most difficult challenge is getting enough pixels in the image to allow us to see the detail when we zoom in on the eye. One example seen on TV would require a 200,000-megapixel camera. That's a lot of detectors to put in a camera sensor! Other key challenges include getting enough light onto each of those detectors, reading all of the data from the sensor (especially at a video rate), building the optics with the necessary quality, and isolating all of the motion so that the image doesn't blur while we image the teensy tiny details from a reasonable distance.

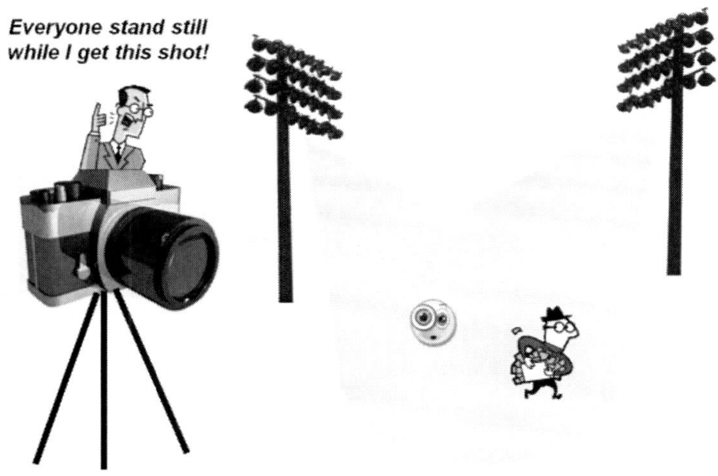

Catching the image of a criminal in the reflection of an eye is going to be a bit of a problem.

Sometimes scientific wizards on a TV show will claim that it is a specialized process, not an incredible camera, that produces the amazing detail when they zoom in. Is there any truth to this magical processing that we see on TV? Perhaps. The process is called super-resolution and attempts to recreate details in a scene that were not captured in the image. We saw that interpolation methods are used to create new gray levels for new pixels when we zoom in on an image, but super-resolution programs go a step farther by making a smart guess as to the detail that would have been in those pixels if the camera had captured them. One way to think of super-resolution processing is that it is a very smart interpolator. Unfortunately, super-resolution processing has shown very limited success at recreating information in a scene. We are a long way off from the thousands-of-percent improvement in resolution needed to make TV crime shows a reality.

Chapter 9
The Way We See It

9.1 You See *What* in that Image?

Once a digital picture has been formed and processed, it's still only a bunch of numbers until we have a computer that displays it for us to see. But, as Hamlet would say, there's the rub. What we see is actually a collection of lit-up dots on the display, but our brain processes these dots to form the image in our head. One person can look at the displayed image and say, "What a cute puppy!" Another person can look at that same collection of lit-up dots and say, "Are you sure that's not a cat?" Thus begins our journey into the interpretation of images.

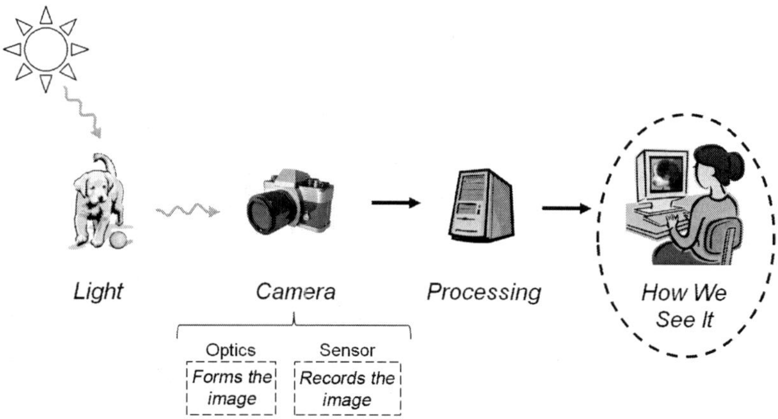

The way we see an image depends on many factors, including the display and our brain.

9.2 A Matter of Interpretation

The human visual system is a complicated beast that is difficult to understand, involving physics, biology, physiology, and psychology. We can think of the human visual system as its own imaging chain, starting with light from the photograph entering our eye (with its own lens and sensor), that light being processed by our brain, and finally imprinting an image in our thoughts of what we think we see. It is this last part that's the kicker because it is based on what each person has learned and experienced up to that point in time, and we can't be sure that two people will interpret an image in the same way. Even when we all agree on what we are seeing, it doesn't mean that any of us perceive the scene as it exists in the real world. Pretty heavy stuff, right?

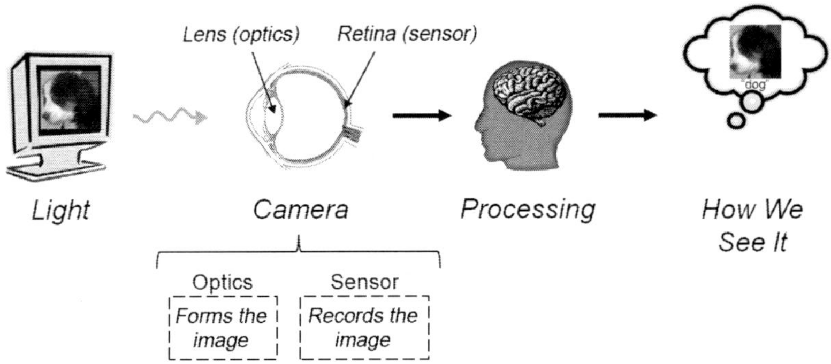

The human visual system is its own imaging chain.

Optical illusions are a great way to demonstrate how our minds can interpret images differently from how they are in reality. Our mind tries to make sense of the information presented to our brain by organizing the pieces into something meaningful. Many of us have memories as children when we gazed up at the sky and saw clouds shaped like bunnies, cars, and sailing ships. If all of the pieces aren't there, our minds fill in the blanks for us. Great examples of this effect can be seen in subject contour illusions, where we fill in the gaps to create edges and shapes where they do not actually exist.

We also rely on our ability to sense brightness differences in a scene to interpret information. Sometimes interpreting a scene requires our minds to sense apparent differences in colors, brightness, and gray shades where none exist. Brightness-constancy illusions illustrate our perception of brightness differences in places where brightness is actually constant. Our minds play this trick on us to make better sense of what we see. It is next to impossible to convince ourselves that the brightness is the same, but a simple check using a photo-editing program proves that they are indeed the same.

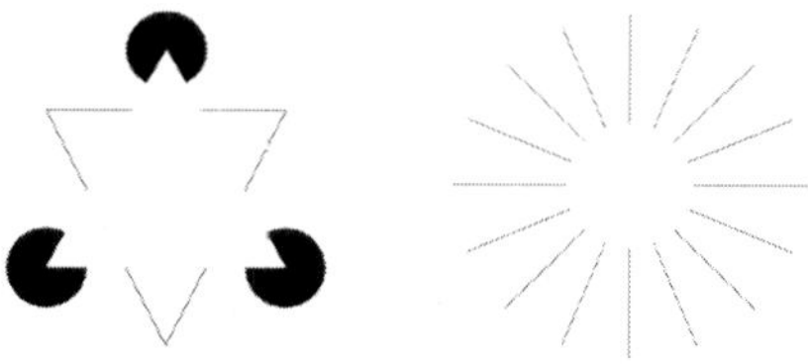

These two illusions demonstrate subject contouring, where our minds fill in missing lines to see shapes that do not actually exist, such as the triangle on the left and the circle on the right.

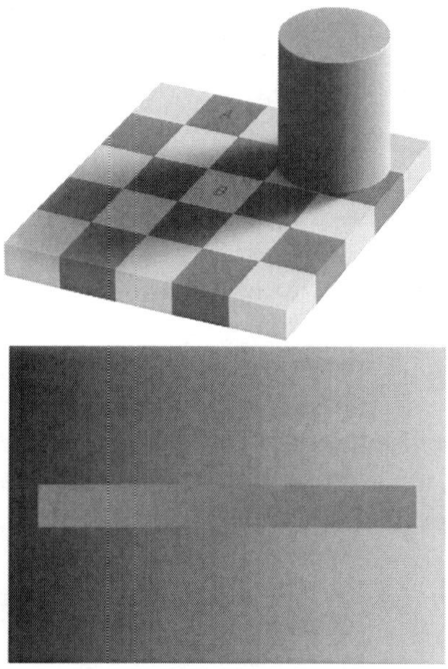

These two illusions demonstrate our minds' inability to perceive the same brightness under certain conditions. The squares labeled A and B in the image on the top are the exact same brightness. The gray strip in the middle of the bottom image is all the same brightness. Believe it or not!

We must always keep in mind that what we see in an image is subject to our minds' interpretation. This happens most often when we see things in an image that we don't understand, and we are tempted to jump to a conclusion that might not be true. One example is an image that has light orbs floating in the scene, interpreted by many people as the spirits of wandering souls, but in reality they are nothing more than light reflecting off dust particles either in the air, on the lens, or in the camera.

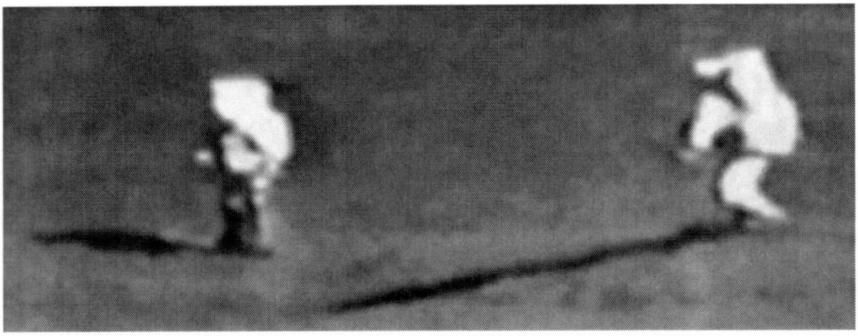

The shadows in this image from the Apollo 11 moon landing give the impression that there must be two suns, but this is an illusion created by capturing a two-dimensional image of the three-dimensional hilly terrain on the moon.

Many misinterpretations result from a loss of depth information in an image. When we look at a single image, we must not forget that we are looking at a two-dimensional image of a three-dimensional world, and this sometimes produces confusing features in the image. Shadows on hilly terrains can appear to point in different directions that give the illusion that there are multiple suns! Another common deception caused by a loss of depth information is the perceived distances between objects in a scene that makes objects look close when they are in fact very far apart. These illusions can be avoided by taking two pictures of the scene from slightly different perspectives to create a stereo pair of images that reveals the three-dimensional aspects of the scene. The photographs on the next page illustrate this deception.

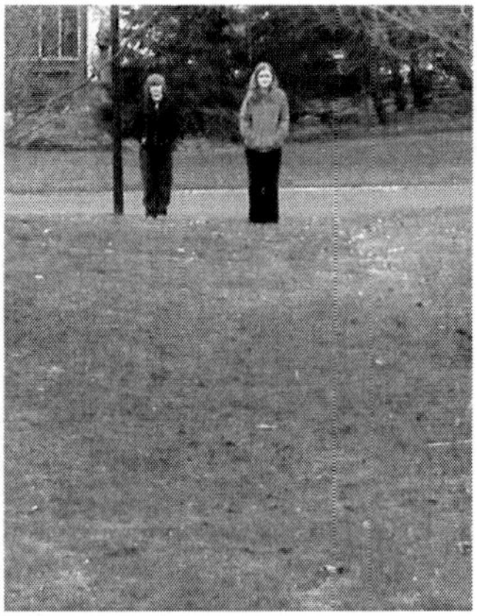

These children appear to be standing next to each other at the curb.

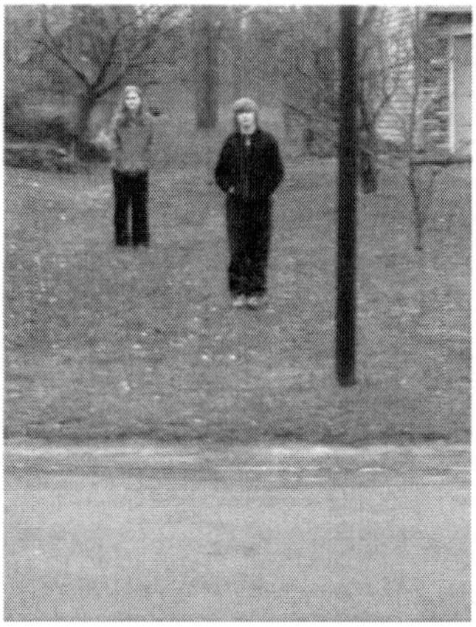

A different perspective shows the truth.

An image is a two-dimensional representation of a three-dimensional world; this can make it hard to see the true distance between objects.

9.3 Camera Designers Need to Keep Our Minds in Mind

The most challenging aspect in designing a digital camera is understanding how the image will be interpreted by our brains when we view images produced from a particular design. A camera is designed to optimize image quality within required constraints, such as size, weight, and cost, but deciding on the best design is difficult without a clear understanding of the relationship between the design factors and the resulting image quality. What happens to image quality if we replace a camera part with one that has a lower cost and lower performance, and does anyone care? Unfortunately, many camera design parameters are dependent on other parameters, so the effect that a design change has on image quality isn't always clear.

Image scientists work closely with psychologists to understand how we perceive and interpret images, so that they can build relationships between the imaging-chain elements and image quality. Studies that relate human perception to presented stimuli are called psychophysical studies and are used extensively by image scientists to understand what drives good image quality.

Psychophysical studies can become overwhelming if every camera parameter, scene variation, and imaging condition is considered, so the studies usually focus on only one or two specific design variables to keep them manageable.

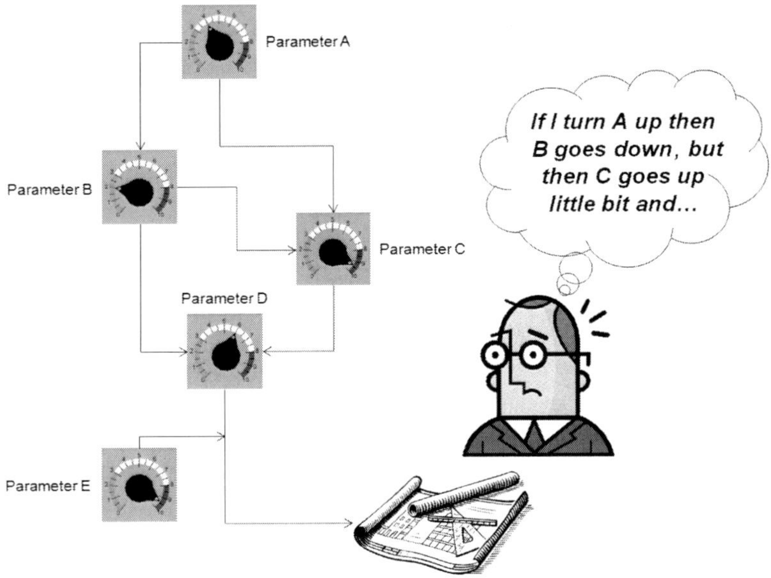

Psychophysical studies help us to understand the relationship between camera design parameters and image quality. Turning one parameter in an attempt to improve image quality can change another parameter that makes the overall quality worse.

It is very important before we start any psychophysical study that we ask the question, "What do you plan to do with the image?" The intended user might want to share vacation memories or perhaps find camouflaged trucks. The images required for either of these uses would need a considerably different camera design when image quality is balanced with other factors, such as size and weight. This is readily apparent when we see professional photographers using large bulky cameras, while teenagers typically only use cameras that fit in their pockets.

Psychophysical studies also help us develop equations that can predict image quality for different camera design parameters. We can use these equations to tweak the camera design, if necessary, without the need for new psychophysical studies. However, developing a general set of equations that relates camera designs to image quality is a very difficult challenge, given the many variables in camera design and the many ways that we take pictures. Once again, the problem resides in the difficulty of writing down equations that accurately describe how our minds interpret images under various scenarios.

9.4 Crummy Viewing Equals Crummy Image

Scientists and engineers can jump through hoops to design a digital camera that takes great pictures, but if the image isn't displayed properly, all of their hair-pulling work is for naught. It is important to have a good-quality display with the appropriate lighting to see the best quality possible. We've probably all experienced the difficulty of trying to view an image on our cameras or phones in the sunlight. Even when we squint while our hand shadows the display, we sometimes find it impossible to see the photo.

The controlled room lighting in our home or office doesn't always provide the best viewing conditions to see all of the detail in an image. Perhaps the most common problem is a bright light, such as sunlight, reflecting off of the monitor while we view the images. It can be surprising how much better the images look if we simply view the images in a windowless room with low ambient diffuse lighting.

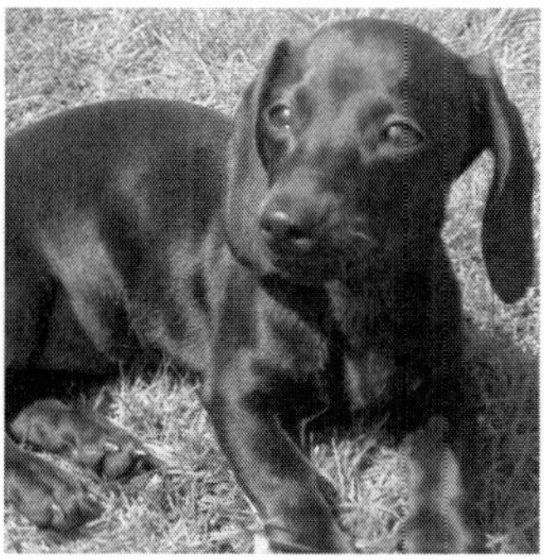

Good ambient lighting conditions

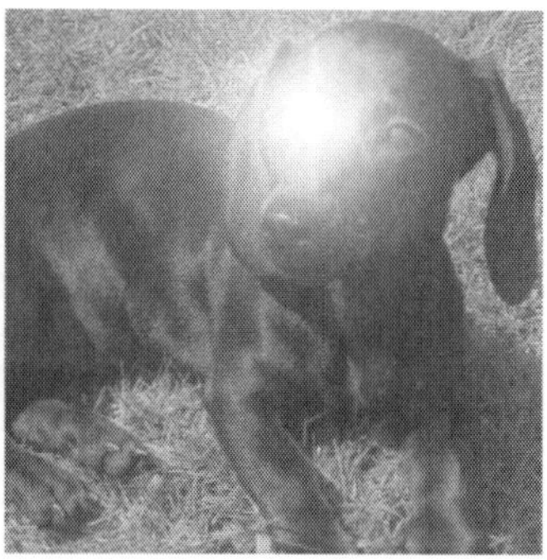

Poor ambient lighting conditions

Many of us spend extra money on a camera that gives good image quality, while at the same time doing nothing about the display conditions to see that quality.

Chapter 10
Afterword: Now What?

Now that we have walked through the imaging chain, what do we do with this information? Most importantly, we understand not only how to make better digital cameras, but also why a digital image is the way it is. We can properly design a digital camera that has the necessary image quality by properly linking the elements of the imaging chain. We can also relate the characteristics of an image to specific elements of a camera's design. This is very helpful when an image has features that we don't like, and we need to change the design to fix it.

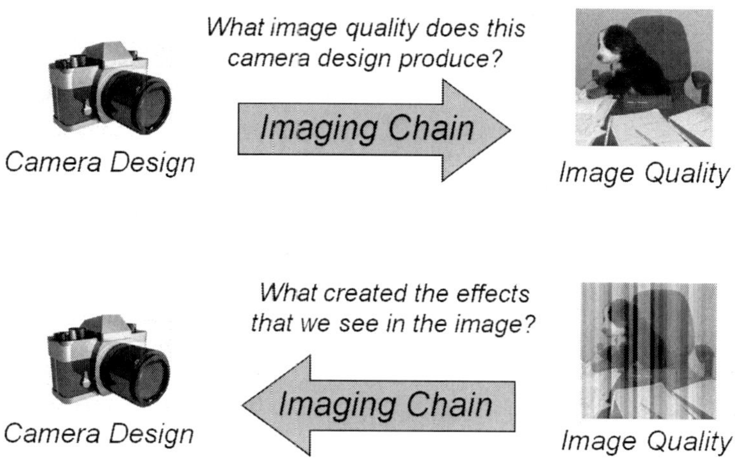

Understanding the imaging chain is critical for understanding not only how a camera design produces the required quality, but also how image features relate back to a camera's design.

Without a proper understanding of the imaging chain, a camera will most likely have elements that are out of balance, resulting in a dominant weak link in the imaging chain. The image-quality effects will be different depending on the location of the weak link, but they will all result in an image with unacceptable image quality.

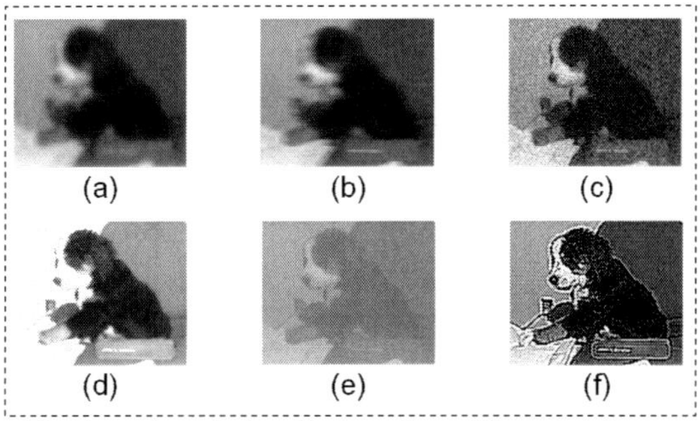

The imaging chains that produced these pictures all have a dominant weak link.

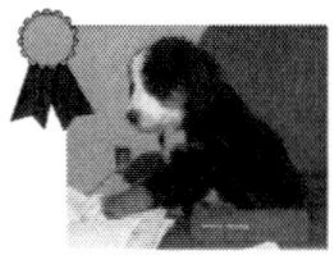

No single dominant weak link in the imaging chain that produced this picture.

Dominant weak links in the imaging chain can be avoided with a proper understanding of the imaging chain. These pictures show weak links by having (a) poor optics, (b) motion blur, (c) sensor noise, (d) overexposure, (e) low contrast, and (f) processing that oversharpened the image.

By mathematically modeling the imaging chain, a computer can produce digital images that are very accurate simulations of pictures a digital camera would produce if we built one. This is an extremely valuable tool, because it allows us to check out designs without spending money on hardware.

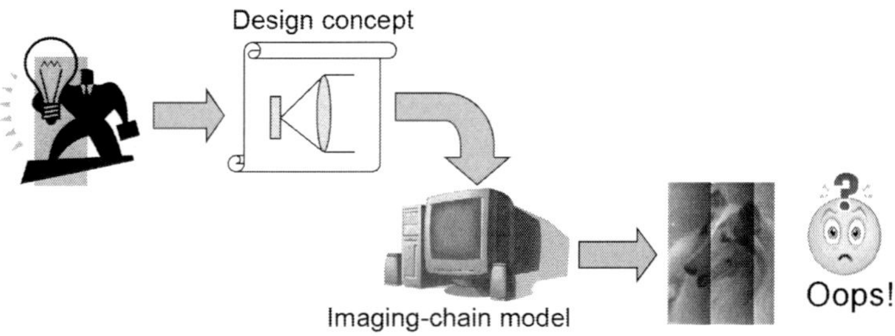

Image simulations can be produced to test a camera design before money is spent (perhaps wasted) on hardware.

Many times, a new camera design is proposed that is either crazy or ingenious, and it's difficult to tell which side of the fence to fall on. Imaging-chain models can be constructed in a computer to test out the idea first to see if it works without breaking the laws of nature. If it passes this test, we can see how well it works under various imaging conditions. More often than not, a clever new idea for a camera will fail this test because it only produces good image quality under a very restrictive range of imaging conditions, such as bright lights or a single color.

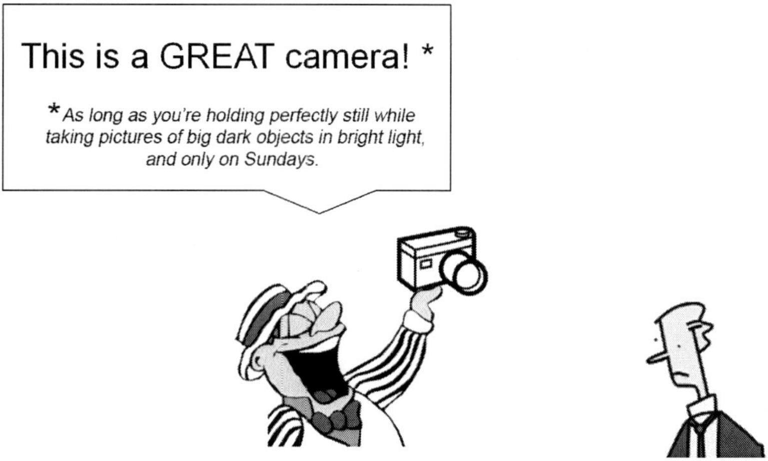

Modeling the imaging chain gives a great understanding of when a camera takes good pictures, but more importantly, when it does not.

So, there you have it. Understanding the imaging chain takes much of the mystery out of creating digital images, but not all of it. We still do not fully understand all of the random events that can influence an image, and we struggle to understand how our minds interpret all aspects of an image. Links in the imaging chain need to be redesigned and reassessed as novel camera concepts are brought forward, but this is good, because it keeps us on our toes. We will continue to build better cameras that meet a wide variety of needs, from pocket cell phones to satellites that whiz far above Earth. And even cameras that take great pictures of puppies, too.

Index

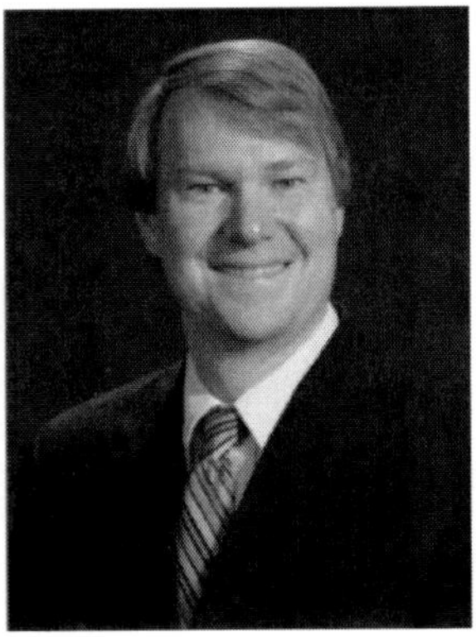

Robert D. Fiete is Chief Technologist at ITT Exelis Geospatial Systems in Rochester, New York. He received his BS in physics and mathematics from Iowa State University, and his MS and PhD in optical sciences from the University of Arizona. In 1987, he joined Eastman Kodak's Federal Systems Division and managed the imaging systems analysis group, responsible for developing and assessing the image quality of novel imaging system designs and advanced processing algorithms. He was an adjunct professor at the Center for Imaging Science at Rochester Institute of Technology, has chaired conferences on imaging science and optics, and has supported the U.S. Department of Justice on cases involving photofakery. He wrote the book *Modeling the Imaging Chain of Digital Cameras* (SPIE Press, 2010), has authored more than 30 technical papers, received 11 patents, and was awarded the Rudolf Kingslake Medal by SPIE.